SEE IT WITH A SMALL TELESCOPE

101
Cosmic Wonders
Including Planets, Moons,
Comets, Galaxies, Nebulae,
Star Clusters, and More

Will Kalif

Ulysses Press

Published in the United States by:
Ulysses Press
P.O. Box 3440
Berkeley, CA 94703
www.ulyssespress.com

ISBN: 978-1-61243-756-9
Library of Congress Control Number: 2017952130

Printed in the United States by United Graphics LLC
10 9 8 7 6 5 4 3 2 1

Acquisitions: Casie Vogel
Managing editor: Claire Chun
Project editor: Claire Sielaff
Editor: Shayna Keyles
Proofreader: Renee Rutledge
Front cover and interior design/layout: what!design @ whatweb.com
Front cover artwork: © Skylines/shutterstock.com
Interior artwork: see page 256

Distributed by Publishers Group West

TABLE OF CONTENTS

THE RICHNESS OF THE NIGHT SKY

The night sky is a deep, rich field of stars. Under normal dark sky conditions, when there is a new moon, there are approximately six thousand objects visible to the naked eye. We often don't realize it, but scores of these objects are not actually stars!

They just appear to be. Galileo made this remarkable discovery when he turned his telescope to the heavens, and you will make this discovery with yours.

With a small telescope, you can observe the wonder of the moon as it orbits a quarter million miles away from us, and things as far away as a spiral galaxy located over 60 million light years away. You can see clusters of stars that formed during the earliest years of our own galaxy, and the large expanses of dust and gas in space that give birth to new stars. There is a profound array of things that you will discover in the night sky. And this book will help you find them.

A TREASURE HUNT THROUGH THE UNIVERSE

We have all browsed the internet and seen many pictures of the planets, galaxies, and the other wonders in the night sky. The Hubble Space Telescope has changed our perspective on the universe, and the images it has taken have set our telescope expectations high. But, you should set your expectations differently when using a small telescope. Your small telescope will not provide you this view of Jupiter, but what you do see will still be remarkable and awe-inspiring. Consider your time with your telescope as more of a treasure-hunting session. How many interesting celestial objects can you find in the night sky? How many different types of objects?

As you use this book, you will take a journey through the night sky. With each object you view, you'll move farther out into the universe and away from our planet Earth. You'll start the adventure in our solar system, exploring remarkable objects including features on the moon, several of the planets, and the moons that orbit around them.

As you move farther away from the solar system, you will explore a vast array of wonders that can be seen within our own galaxy, the Milky Way. Then you will use your telescope to explore the deepest reaches of the universe by finding other galaxies.

Additionally, I have some surprises in store for you. There are some lesser known phenomena in the night sky that are quirky and peculiar, yet no less remarkable. You will finish your tour of the universe by looking at some of these unique objects.

USING THIS BOOK

If you are very new to using a telescope at night, you might not want to jump right into observing the night sky, though it is quite okay to do that. That would make you an adventurous type of person. And adventurous people have been admiring the stars for centuries and using them as guides for their travels and expeditions across both continents and oceans. But I am guessing you are not in need of stars as guideposts for navigation. You just want to see some amazing things with your telescope.

You are, however, going to be hunting for objects that are extremely faint, whose light has taken thousands or even millions of years to arrive at this planet. So there are some things you can do to maximize your viewing potential and get the ultimate ability out of your telescope. For example, it is a good idea to set up your telescope outside an hour before you plan on using it, because its optics— lenses, mirrors, and other viewing mechanisms—

need time to adjust to the changes in temperature and humidity. Doing this will make a big difference in how the telescope performs for you. I recommend that you read Chapter 1 about getting the most out of your telescope.

If you are an adventurer who is already familiar with your telescope and you would really like to get to the fun of the night sky, the next chapter is perfect for you. There, you will find six easy-to-find objects that are also among the best to view in the night sky.

Throughout this book, you will see pictures of objects as seen through a small telescope (has a lens or mirror with a diameter of 6 inches or less). These images, shown in circular views, are actual photographs of the objects. They provide good examples of what you can expect to see with a small telescope on a good viewing night. If you have a medium telescope (has a lens or mirror with a diameter between 7 and 12 inches), you will be surprised at how much more detail you will discern when viewing the objects in this book.

Note that your view may be different when you look through your telescope; there is a lot of variation in what you will see when you look at an object, depending on the size and quality of your telescope, how clear the sky is on that night, and how much light pollution your area of the country has.

DON'T GIVE UP!

I have been an amateur astronomer for several decades. It has been a passion since my teenage years, when I received my first small telescope as a Christmas present. Over the years, I have introduced many people to the wonders of the night sky. But, I have also seen how disorienting the night sky can be at first. It takes a little bit of time and patience to get familiar with what it has to offer.

I encourage you to have a little patience and spend a few nights with your telescope. Once you identify your first few constellations and learn the names of some of the brighter stars, you will quickly get a knack for it, and in no time you will easily recognize many more constellations. From there, you will be able to star hop from one amazing celestial object to another. If you dedicate a few hours to stargazing over the course of a few nights, chances are good that you will become a lifelong lover and observer of the night sky, just as I am.

USING STAR CHARTS

This book uses a uniform set of public domain star constellation charts that were created by a collaboration between the International Astronomical Union and *Sky & Telescope Magazine*. The charts were modified for this book to point out certain objects or features as they are being discussed.

There are a few things you should know when reading a star chart. First, the stars are of different magnitudes, which are represented on the charts as different-sized dots. The larger the dot, the brighter the star. (See more on magnitude on page 10.)

Second, the orientation of star charts can be confusing. The charts are in a book and laid out in a uniform manner, all in one direction. But the night sky is perpetually rotating. Every hour of every night, the orientation of the stars and constellations changes because the Earth is at a different stage of its rotation. When stargazing, you can compensate for the changing orientation by finding the major stars or constellations on the chart, then rotating this book so the chart matches the sky.

CATALOGS OF CELESTIAL OBJECTS

There are a lot of things in the night sky. And for hundreds of years, people have been observing them, cataloging them, and making sense of them. This has created a system by which astronomers and amateur astronomers identify and find them. Throughout this book, I use a few different astronomical terms that you should familiarize yourself with. There's a glossary in the back of the book, which you will find useful, but here are two terms, related to cataloging objects, that you should familiarize yourself with up front.

Many of the deep space objects in the sky were cataloged by an eighteenth-century astronomer named Charles Messier. He created a list of objects we call the Messier catalog. Many of these objects are easy to find and are in this book. Often, they are just referred to by their number, such as M13, which means it is the thirteenth object on the Messier list. All the items in the book that are labeled with an M come from the Messier catalog of deep space objects.

Another abbreviation you will see in this book is NGC. This stands for New General Catalog, which is a catalog of over 7,000 nebulae and star clusters.

It is much more in-depth than the Messier catalog, and while many of the items in this catalog are out of reach of small telescopes, some of them are absolute gems for a small telescope. Those are included in this book.

UNDERSTANDING THE MAGNITUDE SCALE

For each celestial object in this book, I have included the apparent magnitude. This is a measure of how bright a celestial object appears to the human eye here on Earth. It is an important measurement, and you should take note of it when looking for an object. This is because depending on your telescope's size, its type, and the night's viewing conditions, you will have limits on viewing objects at dimmer magnitudes.

Making note of the apparent magnitude of an object when you view it will guide you when selecting other objects in the book. If you have trouble seeing an object through your telescope, or if it is extremely faint, you can select other objects in the book that have a brighter magnitude. They will be easier to find, and to observe.

The scale of magnitude seems a bit counterintuitive. The brighter stars are magnitude 1. Stars that are slightly dimmer are magnitude 2. And as stars get dimmer, their magnitude gets higher in number. Magnitude 6 is the dimmest limit of the human eye. Once an object is higher than magnitude 6, you will need a telescope to see it.

Your telescope, depending on its size, can pick up the light from objects much dimmer than the naked eye can see. A 2-inch telescope can see stars as dim as magnitude 11. And an 8-inch telescope can see objects down to approximately magnitude 14. This example gives you a good sense for why the bigger the telescope, the more you can see with it—the Hubble telescope can see objects as dim as magnitude 30!

CELESTIAL COORDINATES

For each object in this book, I have included its coordinates in the night sky. These coordinates are right ascension (RA) and declination (Dec). They are written like this: 16h 41m 24s, +36° 27' 35.5". The first three numbers are the right ascension and the last three are the declination. This series of coordinates is an internationally recognized system for exactly pinpointing anything in the sky.

They are not needed if you are manually moving a telescope around, but if you have a small telescope that also has the GoTo computerized function, you can punch in these coordinates, and the telescope will bring you right to the object. It makes it very easy for you.

BEST VIEWING MONTHS

The night sky is fluid and ever-moving, and each hour of each night of the month, we see a different perspective. For every object in this book, I have determined the best viewing month or season based on when the object's closest constellation is highest in the sky at 9:00 p.m. Eastern Standard Time.

What this means is that you have a lot of wiggle room for when you can view different constellations and objects. Often, you can extend the viewing months by looking for the constellation shortly before 9:00 p.m., much later in the evening hours, or in the early morning hours. A star wheel (also known as a planisphere) is a great aid with this. You can use it to accurately predict if and when a constellation will be visible.

EASY-TO-FIND CELESTIAL OBJECTS

If you are familiar with your telescope and want to get started right away, here is a list of six objects that you can start out with. They are easy to find and among the best viewing objects you can see.

And these six objects give you a nice tour of the variety of objects viewable in the night sky.

The moon. *Objects 1 through 7 (starting on page 44).* Everybody turns their telescope to the moon first. It is a natural thing to do because it is large, dramatic, and easy to find. For some tips to turn it into a more rewarding experience, see Chapter 4.

Jupiter. *Object 8 (page 54) and Object 9 (page 55).* This gas giant planet is a wonder to observe. It is very bright in the night sky, which makes it very easy to find. And with even the smallest of telescopes, you will be able to see the four major moons that revolve around it. With most small telescopes, you will even be able to see the two major bands on its surface.

Saturn. *Object 10 (page 56) and Object 11 (page 57).* There is simply nothing like this planet. Your first viewing of Saturn through a telescope will be something you will never forget. With a small telescope, you don't get a lot of detail, but you will be able to make out that it has rings. You just have to take a look at Saturn.

M13, globular cluster in Hercules. *Object 21 (page 78).* A globular cluster is a group of thousands of stars packed together in a ball. Sometimes there are even tens of thousands of stars. We will cover many clusters in this book, but if there is one you absolutely must view, it is M13 in Hercules. It is easy to see and relatively bright. On dark nights, it can easily be found with the naked eye.

M42, nebula in Orion. *Object 40 (page 118).* A nebula is a gaseous cloud in space. There are several different categories of nebulae, including dark nebulae, which appear to be black areas in space. In this book, we cover the various types. But, if you are looking for one nebula that is easy to find and dramatic in appearance, then M42 is the one to look at. You will not be disappointed. It is extremely easy to find and worth the look. It may be hard to imagine a gaseous cloud that is millions of light years in size, but that description is true of M42.

M31, the Andromeda Galaxy. *Object 51 (page 142).* We all know what a spiral galaxy looks like. And this one is purely magnificent. It is very large and very bright, as far as galaxies go. On a dark night you can easily find it with the naked eye, which makes it very easy to find with your telescope. Do not miss this beautiful spiral galaxy.

CHAPTER ONE

ABOUT TELESCOPES

A telescope is kind of like an automobile. When you first try it out, everything seems very complicated. But, with a little practice, you will be driving your telescope around the universe with ease.

There are three major types of telescopes: a refractor, which has a lens; a reflector, which has a mirror; and a catadioptric, which has both a lens and a mirror.

When it comes to small telescopes, each type has its advantages and disadvantages. A refractor is easy to use and generally inexpensive. This type of telescope is very recognizable, with its long tube, a lens at one end, and an eyepiece at the other end. The reflector, which has a mirror, is a very economical telescope; it will give you a larger aperture and more light-gathering power for less money. Newtonian and Dobsonian telescopes are both reflectors. The catadioptric telescope is the most expensive, but they will give you the best performance, along with good portability. Catadioptric telescopes are small in size but big in performance.

PARTS OF A TELESCOPE

Every telescope has two important parts. The first is the optical tube assembly (OTA). This is where all the telescope action occurs. The OTA holds the lens or mirror, as well as the eyepiece. When you visualize a telescope, you're probably thinking of the OTA.

The second is the mount, which is what the telescope sits on. The mount is very important because it holds the telescope steady and at a comfortable height, making it easy to use. There are three common types of mounts: the alt-azimuth, equatorial, and Dobsonian.

The alt-azimuth mount is the simplest and easiest to use. It allows your telescope to be moved up, down, left, and right—just like how you would move a camera if it were mounted on a tripod.

Specifically designed for astronomy use, the equatorial mount moves the telescope in a circular motion that mimics the rotation of the earth.

The economical Dobsonian mount is easy to use and made specifically for reflector telescopes. They are popular because they allow users to use a larger reflector telescope while still keeping the cost down.

Most telescopes have an attached counterweight, a heavy weight that balances the weight of the OTA. The weights make the telescope easier to move and point in different directions. If you add anything to your telescope, like a camera, the weight of the OTA will change, and you will need to correct this weight change by sliding the counterweight into a new position.

A LESSON ON EYEPIECES

You might be wondering why eyepieces are measured in millimeters rather than by powers of magnification. This is because an eyepiece might have a different power of magnification if you put it in another telescope, regardless of its size.

Just remember this—the smaller the number on the eyepiece, the higher the magnification. If one eyepiece is marked 4 mm and another eyepiece is marked 10 mm, the 4 mm eyepiece is more powerful.

To calculate the power of your eyepieces, look on your telescope or in the materials that came with it to see what the focal length is. There should be a label. It will say something like F=900 mm or FL=900 mm. This is the focal length of the telescope in millimeters. Divide that number by the focal length of the eyepiece and you will have the power of magnification. So, if the focal length of the eyepiece is 10 mm and the telescope has a focal length of 900 mm, the power of magnification will be 90 times.

WHICH EYEPIECE DO YOU START WITH WHEN OBSERVING?

Telescopes are made so that it's easy to switch out eyepieces without disturbing the telescope. This makes it easy to find an object, locate it with one eyepiece, and then slide in a more powerful eyepiece to get a closer look.

When you are selecting an eyepiece, you should almost always start out with the lowest power eyepiece that you have. That eyepiece will be your workhorse. Low-power eyepieces are easier to use because they show you more of the night sky in their field of view, which makes it easier to find objects. Additionally, they generally gather more light and give you crisper images. They are also less susceptible to atmospheric disturbances such as turbulence, giving you the least amount of distortion in viewing. Once you have located and observed something with your lower power eyepiece, it is quite okay to then switch to a higher power eyepiece.

EYEPIECE FILTERS

An eyepiece filter is a small colored film accessory that threads into the bottom of an eyepiece. They are an excellent way to improve what you can see through the telescope. Various colored filters work better for various objects; they act by filtering out certain bands of light.

You can also get a filter specific for viewing the moon. It filters out some of the bright light that the moon reflects, allowing you to see more detail.

VIEWING OBJECTS THROUGH YOUR TELESCOPE

When viewing through the telescope, what you do with your eyes is important. Here are a couple of "eyeball" tips that you will find very useful. Of course, these tips are all for after you spend at least twenty minutes in the dark so your pupils open up as far as they can go.

First, look through the eyepiece with both eyes open just like experienced amateur astronomers do it. It sounds strange and it might be uncomfortable at first, but this really helps. This is because when you close one eye, the muscles you use to squint cause a slight vibration in the open eye. And for extremely faint objects, that vibration makes a difference. Thus, having both eyes open will help you to see better. Not everyone can get comfortable doing this, but give it a try to see if it works for you.

Second, practice not looking directly at the object in the telescope. Instead, look a little bit to the side of it and see the object from the side of your eye. This brings the light to more sensitive areas of your retina. The rods and cones at the exact center of your retina are less sensitive, particularly to black and white.

Give this technique, called averted vision, a try. You will be surprised how looking a little to the side of the object reveals more details. It is also a very handy technique you can use if you want to do sketches or drawings of celestial objects as you see them through your telescope.

USING THE FINDER SCOPE ON YOUR TELESCOPE

Most telescopes are actually two in one! They usually have a second, smaller telescope mounted on them called the finder scope, which is not very powerful, but is very important. It is a tool that makes finding things in the night sky much easier, because it has a large field of view. You can see a much larger portion of the night sky through it.

Here are two eyepiece views to give you a sense of how useful the finder scope is.

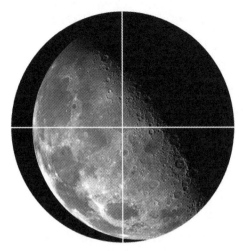

Looking at the moon through the finder scope

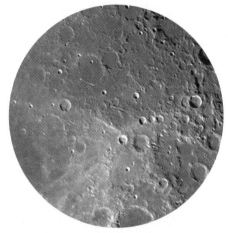

Looking at the moon through the main telescope

In the first image with the crosshairs, you can see almost all of the moon. With this larger viewing area through your finder scope, it is easy to find things in space or on the moon.

But, when looking through the telescope itself, you get a very small, restricted field of view. This is terrific for looking at celestial objects and getting a nice close-up look, but it makes it tricky and challenging to initially find anything. So, to make it much easier, you first find things in your properly aligned finder scope. And then you look through the telescope.

ALIGNING THE FINDER SCOPE

To use the finder scope correctly, it has to be aligned with the main telescope. Aligning is the process of making sure both the finder scope and the main telescope are pointing at exactly the same point in space. This alignment process is easy, and it can be done during the day.

Step 1: Point your telescope at a faraway object here on Earth. It can be anything—the chimney on a house, a stop sign, or a street lamp. For our example, let's assume you are looking at the chimney on a neighboring house. Center the object in your main telescope.

Note: Don't attempt this technique with a celestial object like the moon or a star. This is because celestial objects move as Earth rotates.

Step 2: Without moving the telescope, look through the finder scope. Are the crosshairs in the finder scope exactly centered on the chimney? Probably not!

Step 3: Without moving the telescope, adjust the screw knobs on the finder scope so the cross hairs are centered on the chimney.

Step 4: Look through the main telescope again. Is the chimney still centered in the eyepiece? If not, then your telescope accidentally moved. You probably bumped it! Re-center the image in the telescope and then readjust your finder scope.

Step 5: Magnify the image. Put a more powerful eyepiece on the telescope and check it all again. Readjust the finder scope to keep it nice and centered on the chimney, making sure the telescope is also centered on the chimney. The goal is to get both the main telescope and the finder scope to point at exactly the same spot.

Aligning the finder scope on your telescope only takes a couple of minutes. And it is well worth the effort. People often overlook their finder scope, thinking that it is just a tiny telescope and they won't be able to see much through it. And that is

true; it doesn't provide a detailed view. But, the real usefulness of the finder scope is to help you easily use the big telescope it is attached to.

BEFORE YOUR VIEWING SESSION

Do your research ahead of time. During the day, reference this book, other books, and websites. You should make a plan. Identify a few deep space objects that are easy to find and seem interesting to you. Take into consideration the season you are currently in, the weather, and whether or not there is a full moon. Remember that astronomers dislike the full moon because it washes out a lot of the very dim night sky objects.

Familiarize yourself with your telescope ahead of time. If your telescope is brand new to you or if you haven't used it very much, you should spend some time with it during the day. Point it at faraway objects and try to focus it. If it comes with eyepieces and filters, you should try them all. Try moving the telescope around and pointing it at various things. You want to be very familiar with it, because it can be tricky to use at night in the dark!

Depending on the type of telescope you have, objects may appear upside-down or reversed during the day. This is perfectly normal. Astronomical telescopes behave this way because of the way they are constructed. If your telescope comes with something called an erecting prism, you can use it during the day to view things right-side up. But, there is seldom need to use it at night other than when viewing the moon. And you would use it for the moon only if you need to match up and make sense of lunar charts and maps.

Align your finder scope, if you have one. It is an important tool that will make it much easier for you to find celestial objects. You can do this alignment during the day. Refer to the tutorial on how to align your finder scope.

Take your telescope out early. Let it acclimate to the weather, the temperature, and the humidity. It is an extremely accurate piece of scientific equipment, so small changes can affect it. Take it out an hour before you plan to use it. It will adjust and be much better for use.

When you take your telescope out, place it in a dark location away from lights, and set it on stable ground. You shouldn't put it on a wooden deck, because just the vibrations from your feet will cause difficulty when looking through the eyepiece. Get the mount or tripod on sturdy ground. If the tripod has a holder for the eyepieces, use it to find exactly where your eyepieces are, rather than fumbling around looking for them. That little rack comes in handy and makes viewing much easier.

Be prepared ahead of time. Bring everything out to the telescope before starting. You want to avoid going back into the house for beverages, eyepieces, star charts, a sweater, or anything else. A trip into the house will ruin your night vision. You will see much less and you might think that it is your telescope underperforming.

Dress appropriately. If you are going to be out for a couple of hours with the telescope, be aware that the temperature may drop quite a bit as night falls and deepens. Depending on where you live and how much the temperature may drop, you might want to bring a jacket, extra sweater, or even a scarf out with you to your telescope. It is very difficult to enjoy the subtleties of space when you are cold or uncomfortable. And you want to avoid going back into the house to get extra clothes. That will ruin your night vision.

Give your eyes time to adjust to darkness. This will make a significant difference in what you can see in the telescope. It can take as much as twenty minutes for your pupils to fully dilate. You can start observing right away, but be aware that the ability to observe and locate objects will improve dramatically with a little bit of time in the dark. If you start out your viewing session on one particular deep space object, you might want to return to it an hour later. You will be surprised at how much more detail you will see upon this later viewing. This is both because your eyes will have adjusted and the sky may have gotten darker as the sun moves farther away from the horizon.

TOOL TIP: USE A RED FLASHLIGHT

You are probably going to want to bring a book (preferably this one!) or star chart out with you when using your telescope. But, if you turn on a flashlight to read, you will be ruining your night vision. Get a red flashlight or wrap red cellophane around the lens of a flashlight. Red light has a sharply reduced effect on your night vision.

COMMON TELESCOPE QUESTIONS AND PROBLEMS

Why doesn't the telescope focus? Everything is fuzzy.

Check the lens and the eyepiece for dew. If it is covered in dew, it just needs some time to evaporate. Do not wipe the dew off any of the optical parts of the telescope.

What is the highest power a telescope can go to?

This number can vary significantly depending on the quality of your telescope and the viewing conditions on the night you are using it—although you really shouldn't be concerned with the power of your telescope. For most objects in the night sky, the lower power eyepiece is the best. Between 50 and 100 times magnification is a good range to start with. It is only occasionally that you will try the higher powers.

Why can't I see anything through the telescope?

This commonly happens with beginners. Take the lens cap or cover off the telescope!

Why is the image shaky?

Check the mount or tripod for stability. Make sure all screws are locked tight. And check the ground—did you place your telescope on shaky ground or on a deck? If on a deck, just the vibrations from walking can cause a lot of shake. Move your telescope to a new location. If you have a tripod-style telescope, you can place a weight of about a pound inside a bag and hang it with a hook in the center of the tripod. This will weigh it down and give it more stability.

The telescope moves in weird directions. It doesn't simply go left and right, up and down.

Don't worry! It isn't broken! Some telescope mounts are made that way. It is actually an excellent type of telescope mount called an equatorial mount. It is made to move in the same direction as the night sky. Familiarize yourself with how it wants to move. You will learn how to easily and gently maneuver it with no effort.

It's cold outside! Can I use my telescope inside the house and point it through a window or skylight?

The quick answer to this question is no. Celestial objects are very dim and a telescope is a precise scientific instrument. The glass of a window or skylight, while appearing transparent to the unaided eye, is very distorted when magnified through a telescope.

Why does everything appear upside-down in my telescope?

This is a normal occurrence with many types of astronomical telescopes, and it happens because of the way the optics are built. It helps to save light and make the telescope more powerful. And it doesn't matter when observing most celestial objects. There is no up/down, right/left when it comes to how astronomical objects look. To correct for this directional distortion, however, many telescopes come with an item called an erecting prism. Putting this in the eyepiece holder of your telescope will bring the image to the correct up/down orientation (erecting the image). However, it doesn't correct the right/left inversion.

My telescope doesn't stay where I point it. How do I fix this?

It is simply out of balance. Slide the counterweight up and down to find a spot where the telescope tube doesn't move on its own.

CHAPTER TWO

PHOTOGRAPHY WITH YOUR SMALL TELESCOPE

We have all seen amazing photographs of the moon, planets, galaxies, and other celestial objects. Many of these objects, particularly the deep space objects like galaxies, emit an extremely faint amount of light.

To get good photographs, you need to take exposures that are several hours long. Sometimes, this requires taking hundreds of pictures, then compiling them together with a special software program to make the final image. In short, it takes a lot of work to get those amazing photographs. But, you can get some wonderful pictures with your telescope and some basic equipment. In this chapter, I give you four different ways to do it.

The biggest challenge when it comes to taking space photos is the motion of the Earth. Within a few seconds, any object you are photographing has moved enough to cause a blur in the picture. This blurring is commonly referred to as star trails. Sometimes, these trails are desirable, but not usually. Generally, you can avoid blurring by taking short-exposure photos of less than ten seconds. As a rule of thumb, you should keep your exposures at six to eight seconds.

Determining the right exposure for a photo varies a lot depending on what you are photographing, how bright it is, and the specifics of your camera. The best general advice I can give you is to practice. Trial and error is a good way to learn what works best for the equipment you have and the particular night sky object you are photographing. The more an image is magnified, the shorter time you have to make the exposure. When beginning, use very low magnification in order to get longer exposures that will gather more light.

Your photos will not be as spectacular as the Hubble photographs, but you can take some amazing shots of the moon, the stars, and the planets. For these brighter objects, an exposure of a few seconds will be sufficient to get great pictures.

Vibration in your camera or in your telescope can prevent taking good astrophotos. You can avoid this by using a release cable on your camera or by using the bulb setting (this allows you to keep the shutter open for an extended period of time). In order to avoid shaking the camera when you press the shutter button, cover the lens with a dark material, press the button to open the shutter, pause a few seconds, then remove the dark material covering the lens. Once you have waited the desired exposure time, replace the dark material in front of the lens and then depress the shutter button again, which closes the shutter. This method keeps the camera in darkness any time you press the button.

THE METHODS OF ASTROPHOTOGRAPHY

There are four different methods that you can use to take astrophotos with your telescope. Review these methods to find which one is best for you based on what type of equipment you already have or what kind of photography you want to do.

1. Piggyback astrophotography. Attach your camera to the tube of your telescope, using the telescope simply as a guide for the camera.

2. Afocal astrophotography. Put your camera right at the eyepiece and take pictures. This can be done either by hand or with an inexpensive mounting system.

3. Prime focus. Connect your camera directly to the eyepiece holder of your telescope. There is no camera lens involved. You need a special pair of adapters to securely connect the camera to the telescope. These adapters vary depending on the brand of camera you have but are inexpensive.

4. Eyepiece projection. This is very similar to prime focus in that you connect your camera directly to the eyepiece holder of the telescope. The difference is that there is also an eyepiece between the camera and telescope. For this type of photography, you need camera adapters and you also typically need an extra piece of equipment to hold the eyepiece.

PIGGYBACK ASTROPHOTOGRAPHY

With this type of photography, you mount the camera directly onto the top of the telescope. You use the telescope as a finder and a tracker, but you do not use the optics of the telescope to take any actual photographs.

Some telescopes come equipped with a threaded rod and nut specifically for this purpose. Most modern cameras have a threaded nut on the bottom. You use this to attach the camera to a tripod or other device, like a telescope.

This method is particularly effective if you have a quality DSLR camera that is capable of taking great pictures or has a replaceable lens. But, you can also use it for standard point-and-shoot digital cameras, as long they have the mounting thread.

The photograph here shows a digital point-and-shoot style camera attached piggyback-style to a Newtonian reflector.

If your telescope doesn't come with a camera mount, you can make a mounting apparatus to mount the camera onto your telescope.

This is a great way to begin in astrophotography if you have experience with cameras and photography. What you are doing is simply taking pictures with your camera. The telescope is acting as a mount and a guide.

For this type of photography, you should still consider the rules of exposure mentioned before. Begin by taking photos of the brighter objects like the moon and the major planets, or possibly some of the brighter stars. And you should keep the exposure times low, under ten seconds. If your telescope has tracking or a computerized GoTo system, you can then try to get longer exposures.

You will need to experiment. How well you know your camera will make a big difference in how well you do with this type of astrophotography. If you have large and powerful lenses for your camera, you can take advantage of them with this type of photography.

Note that piggybacking changes the weight and balance of your telescope so you should reset the to bring the telescope into balance. An unbalanced telescope will tend to move at an unstable rate.

AFOCAL ASTROPHOTOGRAPHY

This is an easy option, and it is great for all kinds of cameras, including cellphone cameras and the average point-and-shoot camera. You don't need any special equipment to get some simple pictures of the brighter night sky objects. But, there is equipment you can purchase to make it easier and more reliable.

With this type of photography, you put the camera right at the focal point of the telescope eyepiece. You are effectively using the telescope as a very large second lens on your camera. In this method, you do not remove your camera lens.

The method is easy: You simply place the camera at the eyepiece and take a picture. For the moon, you can do this by hand and also try the video option on your camera. But, for almost everything else, you will need to hold the camera very steady using some kind of a mounting apparatus or adapter that will hold the camera in place at the eyepiece. This type of adapter is called a Universal Digital Camera Adapter (UDCA), which will typically hold any type of camera as long as it has the mounting screw. They also make this type of adapter specifically for cellphones.

A great thing about this type of photography is that you can still adjust the focus of the telescope. The whole apparatus is a good beginner's way to take short-exposure astrophotos without spending too much on equipment. And if your telescope has a GoTo mount, you can typically extend the exposure a bit longer than ten seconds.

Be mindful when snapping the shutter button using this method, because it will cause shake in the camera, so use a bulb cable if possible. If you cannot use a bulb cable, you can use the black hat technique to protect from shake in the image (page 29).

As with piggybacking, attaching the camera to your telescope changes the balance of the telescope. You should adjust the counterweights with the camera in place to compensate for this change.

Universal Digital Camera Adapter (UDCA)

A UDCA is a very simple mechanism that will adjust to just about any type and size of camera, as long as the camera has the 1/4"-20 mounting screw hole. You clamp the adapter to the eyepiece on your telescope. Then, attach the camera to the adapter and adjust it with several thumbscrews on the mount.

And that's it. You are ready to start taking photographs!

PRIME FOCUS AND EYEPIECE PROJECTION PHOTOGRAPHY

These two types of astrophotography are a bit more serious and are used for higher quality astrophotography. You will need a professional style DSLR camera that has the ability to change lenses. The previous methods we looked at use the camera and telescope as separate objects. We simply pair them up in a way that is useful.

With prime focus and eyepiece projection, the camera and the telescope become a single entity in terms of the light path. You use a DSLR camera without any lenses because the telescope becomes the lens. You can take short-exposure pictures with these methods, but they are preferred when getting into longer exposures.

To effectively attach your camera to the eyepiece holder of the telescope, you use a pair of adapters called a T-ring and a T-ring adapter. The T-ring snaps into your camera exactly like any lens would. Make sure you get a style of T-ring that is specific to your make of camera. For example, if you use a Canon camera, you need a T-ring that is specifically for the Canon models.

T-ring and adapter

Next, you need a T-ring adapter. This attaches the T-ring on the camera to the telescope. The type of adapter is dependent on the size of eyepiece holder you have on your telescope. You can see from this picture how it now resembles the shape of an eyepiece.

To attach the camera to the telescope, you simply insert the camera and T-ring assembly into the eyepiece holder of your telescope. Then, tighten the knobs on the holder as you would do with an ordinary eyepiece.

For some telescopes, particularly Dobsonians, this prime focus method might not work because the light path doesn't reach the camera correctly. For a telescope with that characteristic, you would need to use a third adapter so you could screw the telescope eyepiece between the T-ring and T-ring adapter. This is what we refer to as "eyepiece projection." You can use eyepiece projection on all telescopes.

THE BLACK HAT TECHNIQUE OF TAKING ASTROPHOTOS

When taking astrophotos, the biggest problem you run into is the rotation of the Earth. You alleviate that by only taking short-exposure photographs. The second biggest problem you run into is camera and telescope shake. This happens when you press the shutter button on the camera. You can avoid this problem with the black hat technique.

To do it, have your camera and telescope all set up. When everything is ready and the object to be photographed is in the camera and focused, all you have to do is open the shutter and start the exposure. But, don't!

First, place a dark object over the primary opening of the telescope. This darkens everything and doesn't allow any light to get into the camera. Traditionally, photographers used a dark-colored hat for this. But, you can use just about anything that will block out light completely, including a piece of black cardboard or felt.

With the hat in place, now you depress the shutter button to open the shutter. Remove the hat and count out your exposure in seconds. In this case, keep the hat away for eight seconds.

After the eight seconds up, replace the hat. Now you can depress the shutter button to close the shutter. And that's it! You have taken a photograph without allowing any of the shake that can occur when pressing the shutter button.

METHODS FOR TAKING LONGER ASTROPHOTOS

These methods can help you get even better astrophotos.

Stacking photographs. Take a series of photographs, each with an exposure of less than ten seconds. Using a special type of astronomy software called stacking software, you can then stack all your photos together into one photograph that is brighter and more detailed. The software also does a comparison between the photographs to remove any noise or unwanted defects, which is very important. However, for this process to work correctly, you will have to take a series of dark photos using the same telescope and camera setup with the camera lens covered. These are important for the software to help detect noise. You will also have to take another series of pictures called bias frames, which will help the software determine dead and problem sensor pixels in your camera. These pictures are then linked together by the stacking software. By stacking photographs, you can take some amazing astrophotos.

Longer exposures. Some telescopes have a motor attached to them, called a drive motor. It turns the telescope one full revolution every 24 hours and allows you to point the telescope at an object and leave it there. The drive motor will slowly move the telescope and keep the object in view. With a motorized drive on your telescope, you can extend the exposure time when taking photographs. This process is dependent on a few factors, including how good your telescope mount is and how accurate the drive motor is. You can experiment with this to see how well you can do. You may be able to get exposures of 30 to 60 seconds in length.

Tracking photographs. To get extremely long exposures, lasting for minutes or even hours, you need something called an autotracker. It is a webcam that you can connect to your secondary scope or finder scope. You also connect this webcam to a laptop and your telescope tracking mount. The laptop looks through the finder scope and notices any movement of the object. It relays this information to your telescope mount, having it make any adjustments necessary in order to keep the photographed object exactly tracked.

The field of astrophotography is quite amazing, and there are many products that can help you take great deep space photos. There are even dedicated cameras specifically made for the purpose, and cameras that you can connect to your computer so you can observe and even track accurately. Have fun exploring and experimenting!

CHAPTER THREE

NAVIGATING THE NIGHT SKY

When adventuring into the cosmos with your telescope, you are faced with two distinct challenges. You must learn a little bit about the sky itself and how to navigate around it, which can be a bit of a challenge because the location of everything changes from hour to hour as the Earth turns.

But, you also have the second challenge of learning how to use a telescope. Add in the fact that you also do all of this in complete darkness, and you have a recipe for failure! It is very easy to give up, but don't! The advice that follows will help you understand and easily overcome some of the more frequently found obstacles to a successful night under the stars.

IDENTIFYING CONSTELLATIONS

First, let's look at a very easy process for identifying the constellations. And in this book, as with just about every star chart, if you can identify the constellations, you can find most objects in the night sky.

The easiest way to get oriented is to start with one simple and easily recognizable constellation. Once you have found that one, you can simply leapfrog from constellation to constellation. Before you know it, you will be able to identify many areas of the night sky.

The best constellation to start with is Ursa Major, which is commonly known as the Big Dipper. To find the Big Dipper from anywhere in North America, all you have to do is face north. A compass would be helpful for this.

The Big Dipper is made up of seven bright stars in the very recognizable shape of a big ladle. The last two stars in the bowl of the Big Dipper point directly at Polaris, also known as the North Star. Go about five times the distance between those two stars to Polaris. It is as easy as that! And Polaris is the end of the handle of Ursa Minor, also known as the Little Dipper. So in a very small area, you will be able to identify Ursa Major, Ursa Minor, and Polaris. You will be well on your way to scouting around the universe!

Once you find Polaris, you will have found an anchor point for everything in the sky. This is because Polaris is the only celestial object that doesn't appear to move. It is the Pole Star. It stays in the same location over the North Pole, and the sky revolves around it. On future nights, you can look to the same exact spot in the sky to find it.

An asterism is a group of stars that make up a recognizable pattern or shape but is not a constellation. Typically, an asterism is composed of three to seven stars. It is a good term to know and I use it in this book to help you identify objects.

A good example is the Pleiades, which we'll see in Chapter 7. This is a formation of seven stars in an open cluster. It's easy to identify, so much so that it has its own name. But again, it's not a constellation.

Another very famous asterism is called the Summer Triangle. It is composed of three very bright stars, and we will be taking a closer look at it later in this chapter.

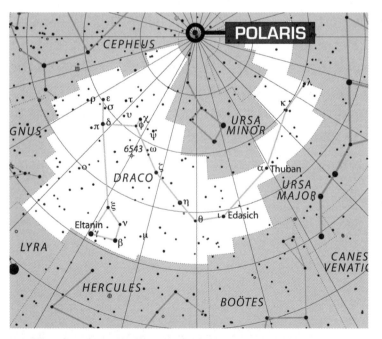

As the Earth rotates, it changes the location of everything in the night sky, except for Polaris, which stays in the same location. Once you have found these first two constellations, you probably have noticed that they are in a different orientation than shown here in this book. To make up for this, simply rotate this book, or any star chart, to the same orientation as what you see in the sky.

With that in mind, let's continue and find another constellation. It is very recognizable and very easy to identify.

If you follow the pointer stars at the front of Ursa Major past Polaris, you will find a W-shaped constellation called Cassiopeia. The North Star is about midway between the two constellations. Cassiopeia is very bright and easy to spot. You have identified yet another constellation!

Once you have found Cassiopeia, you are on the doorstep of the Milky Way galaxy and the constellations of the zodiac. And if you have dark skies, you can easily see the cloudy band of the Milky Way where it starts, right there in Cassiopeia.

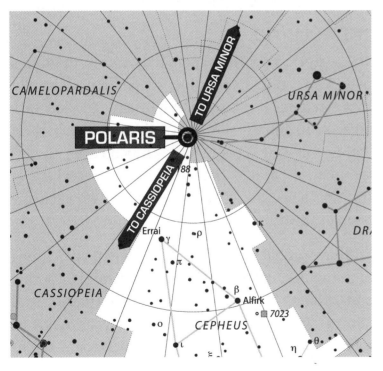

But, before you move on to the Milky Way, you should identify one more constellation, called Cepheus. It is a five-star constellation that looks like a house, located almost between Cassiopeia and the Little Dipper. We saved this one for the fourth constellation because the stars that make it up are not as bright as those of Cassiopeia.

Now, you can identify the Milky Way. It is a dim, cloudy, and irregular band that starts at Cassiopeia, moves past the edge of Cepheus, and on into Cygnus. The constellation of Cygnus is also known as the Swan, and the whole of this constellation is dim. If you are in a city or have light-polluted skies, it may be difficult to see, but there are five bright stars within it that form the shape of a cross. Look for this cross and you have found the major parts of Cygnus. These five stars form an asterism referred to as the Northern Cross.

If the Milky Way is visible to you, it will open up much of the night sky. Follow it past Cygnus to the constellation of Aquila, which is not a prominent constellation but it is extremely easy to find because it contains the bright yellow supergiant star, Altair. The overall shape of the major stars in Aquila form a parallelogram.

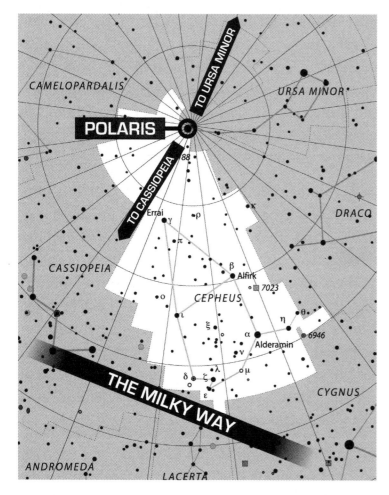

Your journey along the Milky Way has taken you across a large portion of the sky to Sagittarius, which is close to the horizon. The stars that make it up are very bright and form a very recognizable teapot shape. Make note of this teapot because it is a rich portion of the night sky. We will return to it several times in this book.

Once you find Sagittarius, you are at the heart of the Milky Way galaxy, an area that has many deep space objects you can view with your telescope. You are well on your way to identifying all of the night sky that is available where you live. And it all started with finding the Big Dipper. You are more than an adventurer. Now, you are also a navigator.

THE ECLIPTIC

The ecliptic is a thin band that traces the sun's path. You can imagine it like a ring around the sun that stretches out into the night sky. All planetary objects, including the sun and the moon, travel within this ring. This makes it much easier to find the planets because they are not just anywhere in the sky: They are always within the very narrow band of the ecliptic.

You have probably heard of the twelve constellations of the zodiac, including Gemini, Sagittarius, Taurus, and others. The gas giant planets (Jupiter and Saturn) will usually spend most of a year in one or two of these constellations. Mars will roam across multiple constellations over the course of a year.

And Venus will streak through many constellations in the course of a year. The closer a planet is to the sun, the faster it travels, and the farther it roams.

HOW TO FIND THE ECLIPTIC

The sun and the moon move along the ecliptic, just as the planets do. So you can use these easily identifiable celestial bodies to show you where the ecliptic is.

Are you familiar with the area on the horizon where the sun rises and where it sets? Point to the approximate spot where the sun rises. Then, with your pointing finger, raise your arm overhead while moving it toward the location where the sun sets. That arc you have scribed with your arm is an approximation of the ecliptic. Along this line, you will find the planets and the zodiac constellations.

You can do the same thing with the moon. If you are aware of where it rises and sets, you can trace that path to find the bodies in our solar system.

You can also find the ecliptic by using a star chart and finding the visible zodiac constellations. Some of them are extremely easy to find. Once you have identified a few zodiac constellations, it is very easy to expand out to find more of them. Before you know

it, you'll have a good grasp of much of the night sky, because everything is referenced by the constellation it is located in.

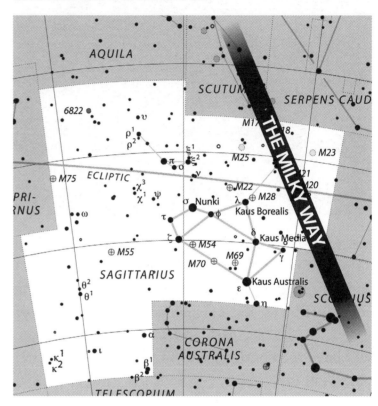

HOW TO FIND YOUR FIRST DEEP SPACE OBJECT

Let's finish off this tour of the night sky by finding a deep sky object through the telescope. The object you will find is Messier Object 52 (M52) in the constellation of Cassiopeia. It is an open star cluster that is bright and easy to find. Once you do, you will have the confidence to tackle any of the other objects in this book.

Choose one of three commonly used methods for finding objects in the night sky.

GoTo. The first method is also the easiest: You can use a computerized controller, called a GoTo, to find objects. If your telescope has a GoTo, it comes with complete instructions on how to use it. You begin the process by inputting various data, including your GPS location, and the current date and time. This tells the GoTo where on the Earth the telescope is located. Next, you orient the telescope by pointing it at two different stars, called alignment stars. These are typically any two bright stars available in the sky. Now you can, with a few keypad strokes, tell the GoTo to "go to" any celestial object in the sky, and the device will automatically move the telescope. GoTos are all very similar in function but do have differences depending on the manufacturer.

If you have a telescope with a GoTo computer and you have it set up, I have some coordinates for you. This is M52 in Cassiopeia: 23h 24.2m 00s, +61° 35' 00".

Star Hopping. The second method is called star hopping. With this technique, you point your telescope at a bright and recognizable star, constellation, or even part of a constellation that is near the desired object. Then, you move the telescope to another star that is even closer to the destination object. You can hop from star to star until you arrive at M52.

This star chart shows you that it is quite easy to star hop to your desired object. You begin by finding Alpha Cassiopeia. From there, you can move the telescope to Beta Cassiopeia. Star hop the same distance from Beta Cassiopeia to M52, moving the telescope to bring M52 into view.

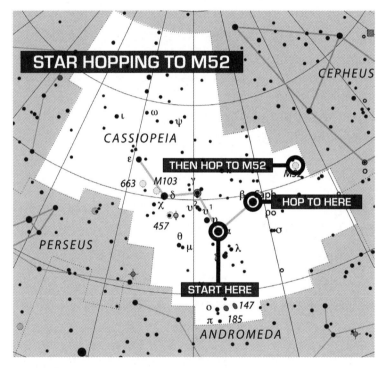

Zooming In. This method is easy and intuitive. It is also an excellent way to find objects that are visible to the eye. If you can see the object with just your eyes, you can successfully use the zoom in method.

Locate the object with your eyes and point the telescope at it. You will not be able to point the telescope exactly at it, but that's okay. Look through the finder scope on your telescope in the direction of the object. Move the telescope so M52 is exactly centered in the cross hairs. The telescope should now also be focused on M52. Using the lowest power eyepiece that you have, move the telescope as needed so the object is centered and in view. Once it is centered and visible, you can zoom in by inserting a higher power eyepiece.

Notice how, without even moving the telescope, the M52 open cluster is going to drift right out of view in less than a minute. You have to move the telescope slightly to keep it in view. This is because the Earth is turning. Take note of the direction you have to nudge the telescope to keep M52 in view because the more you zoom in on it, the faster it will move out of view.

Congratulations! You are officially an amateur astronomer. You are able to use your telescope to find deep space objects and you can zoom in on them to get a closer look.

THE SUMMER TRIANGLE

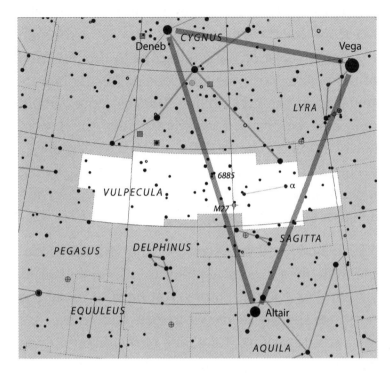

The three stars Vega, Deneb, and Altair are very bright. Finding this asterism gives you a head start on identifying constellations and objects in the night sky: The stars are part of the constellations Lyra, Cygnus, and Aquila, respectively. And you are well on your way to finding many of the 101 objects in this book because thirteen of them are within or around this triangle. These include star clouds, nebulae, a double star, a triple star, a double-double star, and even a meteor shower.

Here are some of the objects in this book that are located within or around the Summer Triangle:

- Object 14, The Milky Way
- Object 15, The Cygnus Star Cloud
- Object 16, The Dark Rift
- Object 36, M11 (The Wild Duck Cluster)
- Object 45, M57 (The Ring Nebula)
- Object 46, M27 (The Dumbbell Nebula)
- Object 49, Barnard's E-shaped Dark Nebula
- Object 50, The Northern Coalsack Nebula
- Object 65, the star Vega
- Object 87, the star Albireo
- Object 89, Omicron Cygni, a triplet of stars

If you are going to continue to learn about the night sky by doing research and getting other books, you will no doubt hear about something called the Summer Triangle. It is an asterism of three bright stars in the summer sky, and during the summer months, it is found almost directly overhead at midnight. This makes it a great starting point for beginners.

- Object 90, Epsilon Lyrae, a double-double star
- Object 92, The Lyrids Meteor Shower

THE WINTER TRIANGLE

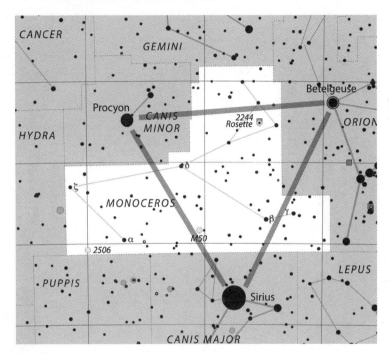

Just like the Summer Triangle, there is a triangle of three bright stars in the winter sky. This one doesn't appear high in the sky. You have to look more toward the south and the horizon during the winter to see it; because of this, it is also called the Southern Triangle.

If available, it is very easy to find this triangle. The three stars that make it up are Procyon in Canis Minor, Sirius in Canis Major, and Betelgeuse in Orion. It is a very fruitful asterism to find, because within and around these three constellations lie many excellent night sky objects you can view with your telescope.

Here are some objects in this book located within or around the Winter Triangle:

- Object 31, M44 (The Beehive Cluster)
- Object 35, the open cluster M35
- Object 38, the open cluster M41
- Object 40, M42 (The Orion Nebula)
- Object 48, M1 (The Crab Nebula)
- Object 63, the star Sirius
- Object 67, the star Rigel
- Object 68, the star Betelgeuse
- Object 69, the star Procyon
- Object 80, the variable star Hind's Crimson Star
- Object 88, Herschel's Wonder Star, a triplet of blue-white giants
- Object 91, The Trapezium
- Object 92, The Orionids Meteor Shower

CHAPTER FOUR

THE MOON

When it is visible, the moon is the first thing we all turn our telescopes to. But, if there is no moon on the night you want to use your telescope, you shouldn't be disappointed, because moonless nights are the best nights to observe everything else in the night sky.

All that light from a full moon causes light pollution that drowns out the dimmer and more elusive objects. Still, observing the moon is an incredibly rewarding experience.

USING A MOON FILTER

Did your telescope come with a moon filter? A moon filter is a small piece that screws onto the bottom of an eyepiece. It works like a pair of sunglasses by reducing the amount of light that comes through to your eye. And when the moon is full or gibbous, it makes a tremendous difference in what you can see on the surface. Make it a point to give the moon filter a try, no matter what phase the moon is in.

1. THE MOON

OBJECT TYPE: Moon

APPARENT MAGNITUDE: -12.74 at full moon

SEASON: Available year-round for most of North America

DIFFICULTY: Easy

The best nights to view the moon are when it is less than full. This is because the terrain features, like craters and mountains, will cast very long shadows, making them very easy to see. Additionally, the reduced light from the moon will have less of an effect on your night vision and your eyes' ability to see fine details. On average, the moon is 238,857 miles away from Earth.

2. LUNAR TERMINATOR

This is the moving line between the dark and the light sides of the moon—the dividing line between night and day. And it is definitely a place you should explore with your telescope. At this point, the sun is low, the shadows are long, and many details are very easy to see. Explore the terminator!

USING YOUR TELESCOPE

Is your telescope showing you an upside-down image of the moon? Don't panic! Learn why this is happening on page 22.

Notice how much more terrain detail you can see in this picture as compared to the larger previous image of the full moon.

LUNAR TERMINATOR

3. THE "WATER" ON THE MOON

If you look at the moon without the aid of a telescope, you can see how the variety of dark and light areas could have been interpreted in ancient times as oceans and seas. We know now that they are not, but the nomenclature we use to identify these large terrain features still stays true to that old thought about bodies of water. We call them *maria*, which is Latin for seas. In reality, they are volcanic plains composed of basalt rock.

I have identified twelve seas and oceans for you. How many can you identify with your telescope?

Take special note of the Sea of Showers. We will be looking at this area later when examining the mountains on the moon.

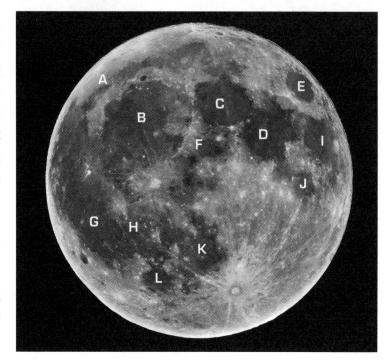

A Sea of Cold, B Sea of Showers, C Sea of Serenity, D Sea of Tranquility, E Sea of Storms, F Sea of Vapors, G Ocean of Storms, H Sea that has become Known, I Sea of Fertility, J Sea of Nectar, K Sea of Clouds, L Sea of Moisture

4. CRATERS

These craters are the result of meteors violently crashing into the moon. And there are thousands of them. But, this doesn't mean that they form every day. The kind of crash that leaves a crater we can see from Earth is an extremely rare event.

This graphic features some of the more prominent craters. How many can you identify with your telescope?

1 Plato, 2 Aristoteles, 3 Eudoxus, 4 Copernicus, 5 Arisatarchus, 6 Kepler, 7 Grimaldi, 8 Langrenus, 9 Stevinus, 10 Byrgius, 11 Tycho

5. TYCHO

Tycho is among the brightest craters on the moon. And it is one of the youngest. This youth is evident by its impact spray, which are light lines that radiate from it. These radiating lines are physically on top of older terrain features. Of course, using the term "young" can be a bit deceiving, because the crater is over 100 million years old.

TIME TO USE YOUR
HIGHER POWERED EYEPIECES

Use your higher powered eyepieces to look around the moon. In particular, take a look at Tycho. It is an excellent example of the varied terrain on the moon and it will reveal plenty of details, depending on the phase of the moon and the angle of the sun. Can you see the peak in the center?

This photograph taken by the NASA Lunar Reconnaissance Orbiter shows a lot of detail in the Tycho crater. Notice how the central peak is very visible because of the long shadows.

6. MOUNTAINS ON THE MOON

Mountains on Earth are formed by tectonic movement and volcanic eruptions. The moon doesn't have these geological phenomena. Mountains on the moon have been formed by the impact of meteors and their melting effect on the lunar surface. With a small telescope, it can be challenging to find individual mountains. But, there are three major mountain ranges that are easily found and identified. They wrap around the eastern edge of the Sea of Showers.

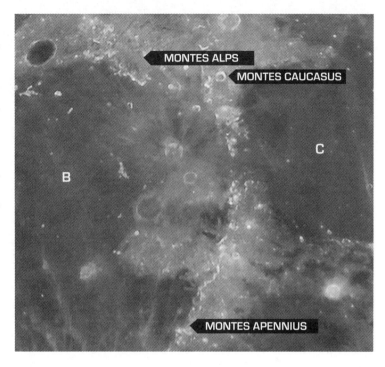

B Sea of Showers, C Sea of Serenity

7. THE LANDING SITE OF APOLLO 11

While it isn't possible to see the lunar lander on the moon with any telescope, you can look at the general area where Apollo 11 landed and the first men walked on the moon. Look just south of the Sea of Tranquility for the approximate area. The X in the image shows you the site.

APOLLO 11 SITE ▶ X

D Sea of Tranquility

CHAPTER FIVE

WITHIN OUR SOLAR SYSTEM

Our solar system is a lively place with many extraordinary things that you can observe with your telescope. Many objects in the night sky are typically very easy to find because they stay in the same place, appearing to rotate in position every 24 hours as the Earth turns.

The planets, however, are wanderers. They do not stay in the same place over the course of a year, but change their location against the background of the stars. This wandering is how they came to be called planets. The ancient Greeks called them *planetes asteres*, or wandering stars. To view them, you have to track them down, but nowadays this is pretty easy to do.

In this chapter you will locate and view Jupiter, Saturn, Mars, and Venus with your telescope. They are all very easy to find because they are visible to the naked eye.

Remember: When using your telescope, start with your lowest power eyepiece. You can easily take that eyepiece out and put a more powerful one in later.

PLANETARY CHARTS AND STAR CHARTS

A star chart shows you constellations and stars in the sky. A planetary chart shows you the location of a planet against the background of the stars.

One of the handiest things about a star chart is that it also easily shows you the brightness of individual stars. The bigger the dot, the brighter the star.

To find these four major planets, a star chart is a nice aid, though you can find them without one. A good rule of thumb is to look through the band of the ecliptic for a bright star that doesn't twinkle. Currents in the Earth's atmosphere will cause stars to twinkle, but not planets. So, if you don't want to reference planetary location charts, this is one way to find some of the planets. This method is relatively easy because the planets are among the brightest objects in the night sky.

TIPS FOR VIEWING PLANETS IN YOUR TELESCOPE

When viewing planets, you want all the same viewing conditions as those you'd seek with deep space objects, such as a nice dark sky and eyes that are well adapted to dark. But the planets are much brighter than galaxies or star clusters, so it's okay if you're gazing on a brighter night.

However, planets do have a significant width. They are not pinpoints like the stars are, which is also the reason why they don't twinkle. The larger appearance of the planets means that the atmosphere can have a very dramatic effect on your observations. These effects are amplified when planets are low on the horizon, because the light from a planet has to pass through a significantly deeper slice of atmosphere. Consider this when planning an evening of viewing. Check your planisphere or star charts for the location of the constellation throughout your observing session. If the constellation is low on the horizon, you will get a lot of distortion when viewing the planet.

A BONUS CHALLENGE

If you have found and observed the big four planets that are very easy to find, you may want to give two of the more elusive planets, Uranus and Mercury, a try. They can be seen with a small telescope. But, Mercury is difficult because it is so close to the sun, and Uranus is a challenge because it can't first be located with the naked eye. You will need planetary charts to find these two planets. There are many excellent online resources that will help you find them, including Stellarium.org, a great desktop program that is free and will show the exact location of any celestial object for any time or day of the year.

8. JUPITER

OBJECT TYPE: Planet

MOONS: Jupiter has 69 known moons. Europa, Ganymede, Io, and Callisto are easily seen with a small telescope.

APPARENT MAGNITUDE: Between -2.94 and -1.6

SEASON: Available majority of the year for most of North America

DIFFICULTY: Easy

HOW TO FIND JUPITER

Locate Antares, the largest and brightest star in Scorpius, and then locate the five major stars of Libra. During the year, the path of Jupiter crosses these two constellations from right to left.

Jupiter is the easiest of all the planets to find because it is very bright and is often high in the sky. Use the rules of thumb for finding it. Do you see any very bright objects in the sky that do not twinkle? Is it in the plane of the ecliptic? Turn your telescope to it and give it a look.

The view of Jupiter with any telescope is a pleasant surprise. Start with the lowest power eyepiece you have. You will get a sense for the viewing conditions. If it looks crisp and sharp, you can move to a more powerful eyepiece magnification.

With most small telescopes, and a reasonably good viewing night, you should be able to make out the two major bands that cross the surface of Jupiter.

9. THE MOONS OF JUPITER

The four major moons of Jupiter (Europa, Ganymede, Io, and Callisto) are called the Jovian satellites, and they are very easily seen in a small telescope. The eyepiece view of Jupiter shows them arranged in a band to the left and right of the planet. One of the remarkable things about these moons is how quickly they move. View the gas giant planet a few hours later, or on the next night, and they have all moved. Or you may not see all four. One or more of them may be in front of or behind Jupiter.

A fun activity is to draw out the location of the four moons in relation to Jupiter. And a few hours later, or on another night you can draw another sketch. This will be your own personal catalog of how the moons of Jupiter revolve around the gas giant.

10. SATURN AND ITS RINGS

OBJECT TYPE: Planet

MOONS: Saturn has 53 named moons. Titan is easily visible in a small telescope.

APPARENT MAGNITUDE: varies between -0.24 and +1.47

SEASON: Available majority of the year for most of North America

DIFFICULTY: Easy

This is truly one of the most remarkable objects in the night sky. With a small telescope and on an average viewing night, you might not be able to get a sharp and crisp view of the rings. They may look like blobs on each side of the planet. Try switching to a higher power eyepiece to get a closer look.

When looking through your telescope, can you make out the major division between the inner and outer rings? This is called the Cassini Division.

As Saturn travels around the sun, our view of the planet changes. The rings appear to tilt, with the angle changing over a fourteen- to fifteen-year cycle. During some years, the rings face us edge-on, so they seem to disappear from view.

11. TITAN

OBJECT TYPE: Moon

APPARENT MAGNITUDE: +8.2 to +9.0

SEASON: Available majority of the year in most of North America

DIFFICULTY: Easy

Titan is Saturn's largest moon. It has a sixteen-day orbital period around Saturn, so it is very easy to see how its position changes from night to night. And it is very bright. You can easily mistake it for a star. But, there generally aren't any stars within the same telescope view of Saturn, so if you see a bright object next to the planet, the chances are extremely good it is Titan.

A great activity you can do is to draw a sketch of Saturn and Titan. Then on following evenings, you can make additional drawings showing how Titan has changed its position.

12. MARS

OBJECT TYPE: Planet

MOONS: Phobos and Deimos

APPARENT MAGNITUDE: -3.0 to -1.4

SEASON: Available majority of the year in most of North America

DIFFICULTY: Easy

HOW TO FIND MARS

Mars is bright red and easy to identify with the naked eye. Unlike a star, it will not twinkle; also unlike surrounding stars, you should notice it expanding into a small disk with some magnification.

Mars is visible with the naked eye and easy to identify because of its red color. But, it is much smaller than the gas giants Jupiter and Saturn, so a small telescope will not reveal much in terms of detail. Instead, it will look like a small red disk. Occasionally, with excellent dark and calm skies, you may be able to catch some detail.

If your telescope comes with some color filters, try them when viewing Mars. They can help to resolve more detail. A light blue filter is often very effective in bringing out details in the Martian polar ice caps. Start out with a low-power eyepiece just to orient yourself. Then move to a higher power. If the sky is very clear and calm, you could get some details like variations in dark and light on the surface or the brightness of the white polar caps.

13. VENUS

OBJECT TYPE: Planet

MOONS: None

APPARENT MAGNITUDE: -4.9 and -3.8

SEASON: Available majority of the year for most of North America

DIFFICULTY: Medium

Venus is extremely bright so you can't mistake it when it is visible. But, because it is closer to the sun than the Earth is, it is never high in the sky. It can typically only be seen shortly after sunset or shortly before sunrise, usually near the horizon. So, you have to look to the ecliptic during these times. Look for a very bright star that doesn't twinkle. Being so bright and near the horizon makes Venus susceptible to the misconception that it is an unidentified flying object!

Using your telescope on Venus will reveal a disk, and sometimes it will display itself in phases just like the moon does. This change in appearance is worth the viewing alone. And if you are keeping a sketchbook of your observations, be sure to find Venus and draw its current phase.

The telescope view here reveals Venus in a slightly gibbous phase. There is no planetary chart for Venus because of how short its orbit is around the sun.

CHAPTER SIX

THE MILKY WAY GALAXY

Our solar system is a small group of bodies that lie within a much larger body, a spiral galaxy called the Milky Way. It is composed of millions of stars slowly swirling around its central core.

This picture here shows you what a spiral galaxy looks like. You can't see the Milky Way galaxy from this perspective because you are inside it! But, being inside it means you are relatively close to a whole lot of amazing celestial objects.

We see the Milky Way as all the stars and objects in the sky. And this includes many things other than stars, like gaseous clouds and tightly bound clusters of stars. Along the midsection of the Milky Way, there is a higher concentration of these celestial objects and gases. This core forms a band in the night sky. You can see it very easily on dark nights, and you can explore it with a small telescope.

STAR CLOUDS IN THE MILKY WAY

Near the center of the Milky Way galaxy, you'll find some very dense areas known as star clouds, which are collections of millions of stars in relatively small areas. Many of them are so dramatic that they have their own names. You will take a look at some of the more famous ones. This is a great way to familiarize yourself with how to find things in the night sky, because for most of these star clouds, you can identify them first with just your eyes. Then you can turn your telescope to them for more detailed examination and exploration.

Next you will travel along the Milky Way from west to east and look at some of the larger and more interesting star clouds in the core of the galaxy, ending with the general area where the absolute center of the galaxy is.

You started your adventure by identifying Cassiopeia as the place where the Milky Way begins. From there, you traveled east to the constellation Cygnus, which is your first star cloud to examine.

14. THE MILKY WAY

OBJECT TYPE: Spiral galaxy

APPARENT MAGNITUDE: Varies

COORDINATES: 17h 45m 40.04s, -29° 00' 28.10"

CONSTELLATION: Spans a wide area of sky; the central area of the Milky Way is in Sagittarius

SEASON: March through October

DIFFICULTY: Medium

Technically, you can't look at the Milky Way with your telescope; not as a whole, anyway. It would be like trying to look at an elephant with a microscope. It is simply too large. But, once you have found the Milky Way as it stretches across the sky, you can explore parts of it and look at various interesting areas with your telescope. As you go through this chapter, you'll explore the Milky Way from west to east.

Due to its size, you don't need a star chart to find the Milky Way (but I have provided one, just in case). Depending on the season and the time of night, it can span much of the night sky. However, you do need very dark skies with minimal light pollution and your eyes have to be dark adjusted. Spend at least fifteen minutes in the dark under a clear sky.

The Milky Way will reveal itself as a wide and wispy trail with a variety of densities. It will appear almost like a very thin network of cotton balls or spider webbing.

The Milky Way galaxy crosses several constellations. Once located, point your telescope anywhere within the galaxy's length. Take your time and have some fun. There are lots of hidden diamonds within this large expanse.

Notice how the Milky Way starts out very light and wispy at Cassiopeia. As you move east, toward Cygnus, it gets denser because you are approaching the center of the galaxy. You might want to start your telescope exploration on the east side. There is simply more stuff there!

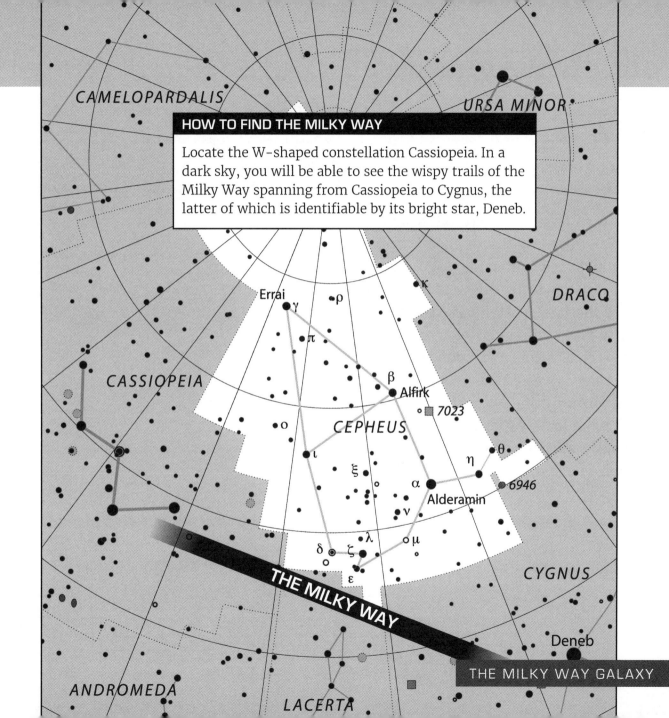

HOW TO FIND THE MILKY WAY

Locate the W–shaped constellation Cassiopeia. In a dark sky, you will be able to see the wispy trails of the Milky Way spanning from Cassiopeia to Cygnus, the latter of which is identifiable by its bright star, Deneb.

CAMELOPARDALIS

URSA MINOR

DRACO

Errai
γ
ρ
κ
π
CASSIOPEIA
β
Alfirk
7023
ο
CEPHEUS
θ
ι
η
ξ
ο
α
6946
Alderamin
ν
δ
ζ
λ
ο μ
ε
CYGNUS

THE MILKY WAY

Deneb

ANDROMEDA

LACERTA

15. THE CYGNUS STAR CLOUD

OBJECT TYPE: Star cloud

CONSTELLATION: Cygnus

APPARENT MAGNITUDE: Varies

COORDINATES: Spans a very wide area

SEASON: June through October

DIFFICULTY: Easy

This cloud is very large and very interesting to explore, especially the bright area that is east of Deneb and south of Vega. Within the star cloud, you will see a lot of variation in densities and brightness. There are areas that are densely packed with stars and areas that are dark with interstellar dust. There are even some secrets within and around the cloud, like the North America Nebula (NGC 7000), which is a large nebula formed from ionized hydrogen gas; and M29, which is a loosely formed open cluster of stars. These objects are not listed in this book; enjoy finding them on your own!

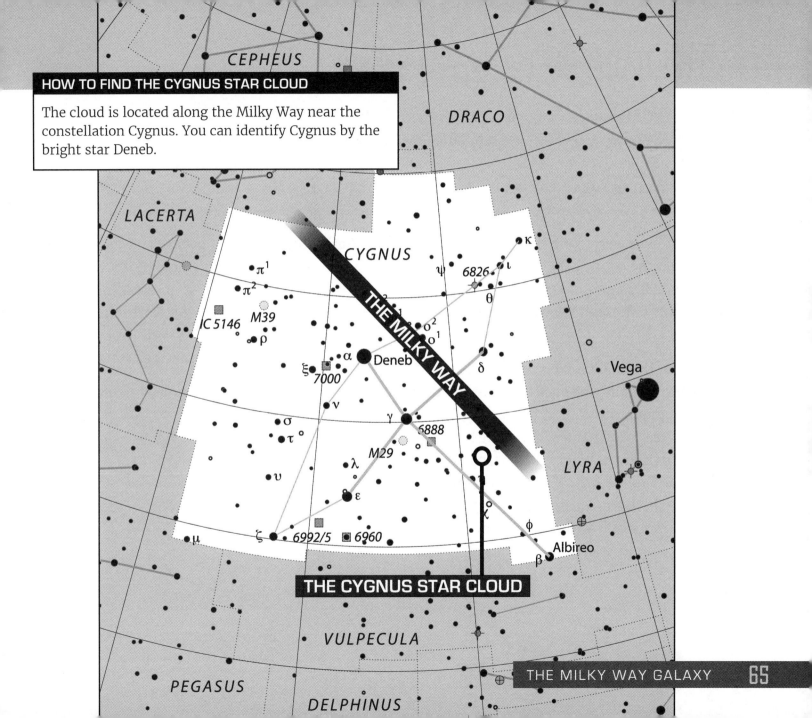

HOW TO FIND THE CYGNUS STAR CLOUD

The cloud is located along the Milky Way near the constellation Cygnus. You can identify Cygnus by the bright star Deneb.

THE CYGNUS STAR CLOUD

16. THE DARK RIFT

OBJECT TYPE: Star cloud

CONSTELLATION: Cygnus, Aquila, Ophiuchus, Sagittarius, and Centaurus

APPARENT MAGNITUDE: N/A

COORDINATES: Spans a very wide area

SEASON: June through October

DIFFICULTY: Easy

The Dark Rift is a very long and broad area of darkness within the band of the Milky Way. It starts at around the star Deneb in Cygnus and extends a very long distance through Cygnus, and all the way to Centaurus.

The darkness of this star cloud is not the absence of matter. Rather, it is the presence of molecular dust clouds. Throughout the Dark Rift, new stars are being born. It has many names, including the Great Rift and the Dark River. Where it overlaps Cygnus, it is called the Cygnus Rift. This northern portion within Cygnus is the most dramatic, but you can explore much more of it with just your eyes, a pair of binoculars, or your telescope.

The Dark Rift is often overlooked by amateur astronomers because it is thought to be empty. However, there are a lot of interesting celestial objects within it, including the Northern Coalsack Nebula (page 138) and the Veil Nebula (NGC 6960), which is a cloud of ionized gas and dust.

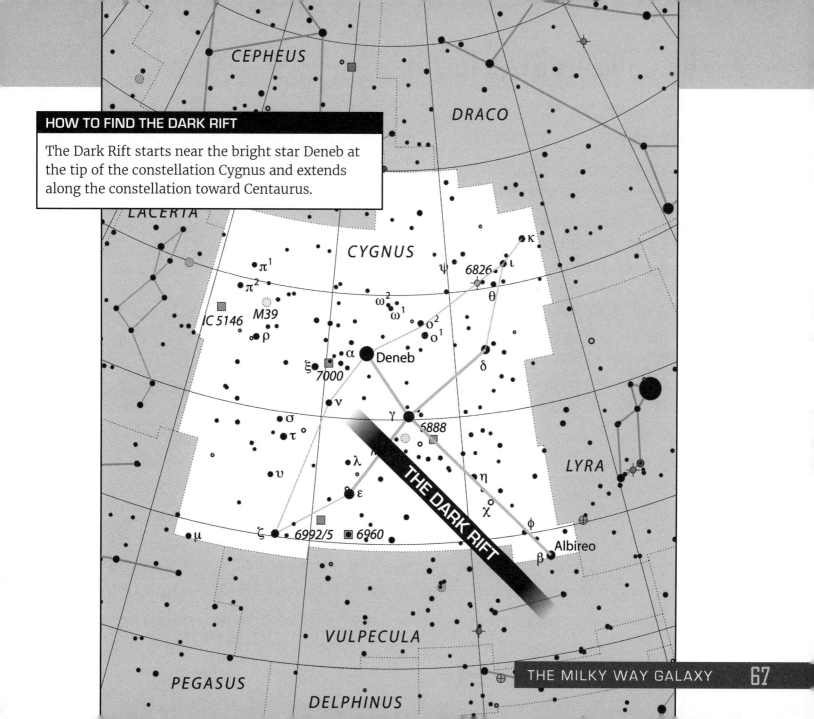

HOW TO FIND THE DARK RIFT

The Dark Rift starts near the bright star Deneb at the tip of the constellation Cygnus and extends along the constellation toward Centaurus.

CEPHEUS

DRACO

LACERTA

CYGNUS

π^1

π^2

κ

ψ 6826 ι

ω^2 θ

ω^1

IC 5146   M39 o^2

ρ o^1

α Deneb δ

ξ 7000

ν

σ γ

τ 6888

LYRA

λ

η

υ

χ

ε

ϕ

μ ζ 6992/5   6960   Albireo

β

THE DARK RIFT

VULPECULA

PEGASUS

DELPHINUS

17. THE SCUTUM STAR CLOUD

OBJECT TYPE: Star cloud

CONSTELLATION: Scutum

APPARENT MAGNITUDE: N/A

COORDINATES: Spans a very wide area

SEASON: August through October

DIFFICULTY: Easy

The constellation of Scutum is one of the smallest constellations in the sky. It consists of only four stars. And this star cloud engulfs all four. The Scutum star cloud has been called the gem of the Milky Way, and for good reason. It is rich with a very high density of stars, and within it, you can also find the open star cluster M26 and The Wild Duck Cluster (M11, page 108). Time exploring this star cloud will be time well spent.

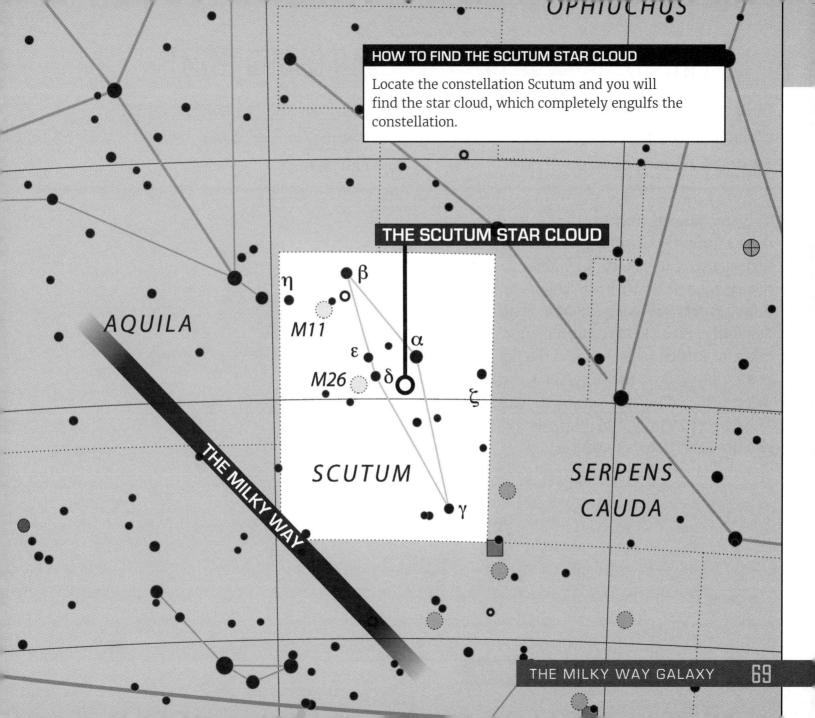

OPHIUCHUS

HOW TO FIND THE SCUTUM STAR CLOUD

Locate the constellation Scutum and you will find the star cloud, which completely engulfs the constellation.

THE SCUTUM STAR CLOUD

AQUILA

η

β

M11

ε

α

M26

δ

ζ

THE MILKY WAY

SCUTUM

γ

SERPENS
CAUDA

18. THE SMALL SAGITTARIUS STAR CLOUD (M24)

OBJECT TYPE: Star cloud

CONSTELLATION: Sagittarius

APPARENT MAGNITUDE: +4.6

COORDINATES: 18h 17m 00s, -18° 29' 00"

SEASON: June through August

DIFFICULTY: Easy

This star cloud is truly wonderful. Its magnificence comes from the fact that it is both very bright and very dense with objects of interest. It is also an oddity on the Messier list of objects because all the other objects are compact and small. The Small Sagittarius Star Cloud is much more dispersed and takes up a much larger space in the sky. Within this star cloud, you will find several interesting celestial objects, including two open star clusters on the Messier list (M18 and M25), several dark nebulae, and the Swan Nebula (page 122).

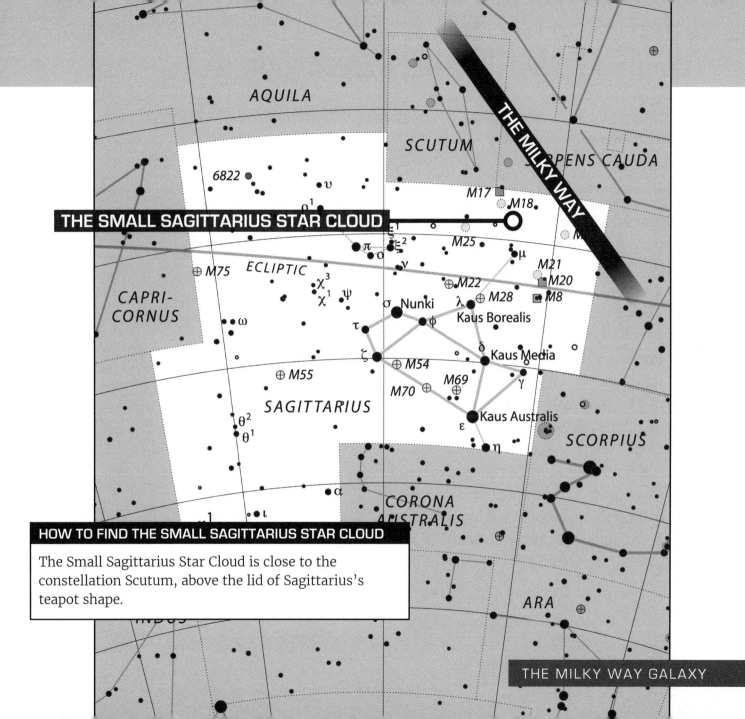

THE SMALL SAGITTARIUS STAR CLOUD

HOW TO FIND THE SMALL SAGITTARIUS STAR CLOUD

The Small Sagittarius Star Cloud is close to the constellation Scutum, above the lid of Sagittarius's teapot shape.

19. THE GREAT SAGITTARIUS STAR CLOUD

OBJECT TYPE: Star cloud

CONSTELLATION: Sagittarius

APPARENT MAGNITUDE: N/A

COORDINATES: Spans a very large area

SEASON: June through August

DIFFICULTY: Easy

The center of the Milky Way galaxy is located in the constellation of Sagittarius, and from that information, we can assume that there are a lot of interesting objects to see in this constellation. Among these interesting objects is this most spectacular space cloud. It is very easy to find and is well worth turning your telescope to. It covers an extremely large area of the night sky, so you can't see the whole thing in the eyepiece of your telescope. You should locate it and look around it with your telescope, starting out with your lowest power eyepiece and moving to higher powers once you get familiar with its arrangement and shape. Within it, you will find a wide variety of interesting celestial objects, including the open star cluster M21 and the Lagoon Nebula (page 120).

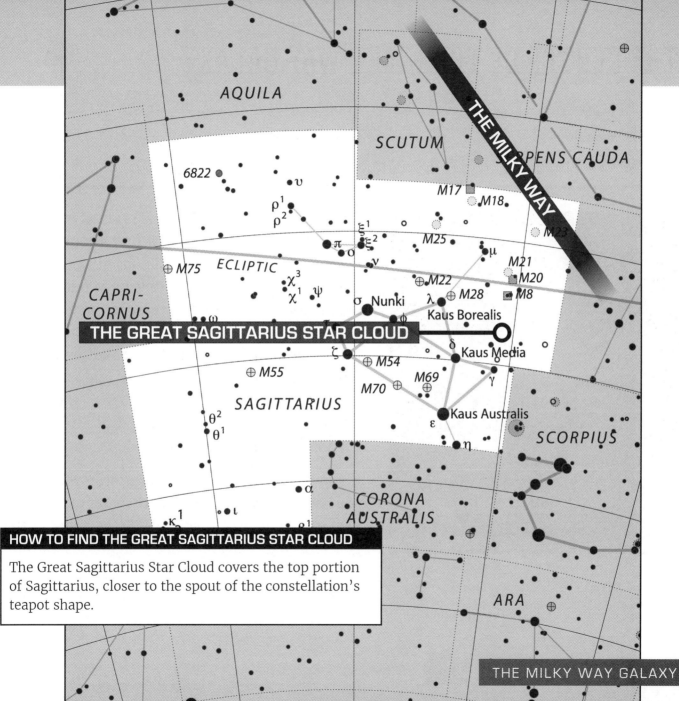

AQUILA

SCUTUM

SERPENS CAUDA

THE MILKY WAY

6822

υ

ρ¹
ρ²

π
ξ¹
ξ²
σ
ν

M17
M18
M23

M25
μ
M21
M20
M8

M75

ECLIPTIC

χ³
χ¹
ψ

M22
M28

λ

CAPRI-
CORNUS

ω

σ Nunki
φ

Kaus Borealis

δ

THE GREAT SAGITTARIUS STAR CLOUD ●——○

Kaus Media

ζ
M54
M55

M69
M70

γ

SAGITTARIUS

θ²
θ¹

Kaus Australis

ε
η

SCORPIUS

α

ι

CORONA
AUSTRALIS

κ²

ρ¹

ARA

HOW TO FIND THE GREAT SAGITTARIUS STAR CLOUD

The Great Sagittarius Star Cloud covers the top portion
of Sagittarius, closer to the spout of the constellation's
teapot shape.

20. THE CENTER OF THE MILKY WAY GALAXY

OBJECT TYPE: Spiral galaxy

CONSTELLATION: Sagittarius

APPARENT MAGNITUDE: N/A

COORDINATES: 17h 45m 40.04s, -29° 00' 28.10"

SEASON: June through August

DIFFICULTY: Currently impossible

The center of the Milky Way galaxy is an interesting and active place. There, you'll find a supermassive black hole, surrounded by millions of stars. We can see this general area with a telescope but we can't see the exact center for two reasons. First, the black hole swallows everything, including light. Because no light is emitted, we can't see anything with our telescopes. Second, there is a lot of interstellar dust and gas between us and the center of the galaxy. This dust and gas obscures our view, so we can only get a look in the general area.

If you have found the Great Sagittarius Star Cloud, you have found the approximate center of our Milky Way galaxy. It is defined by NASA as being in this generalized area in and around Sagittarius. Infrared photography, which can see through a lot of the space dust and clouds, was used to determine this.

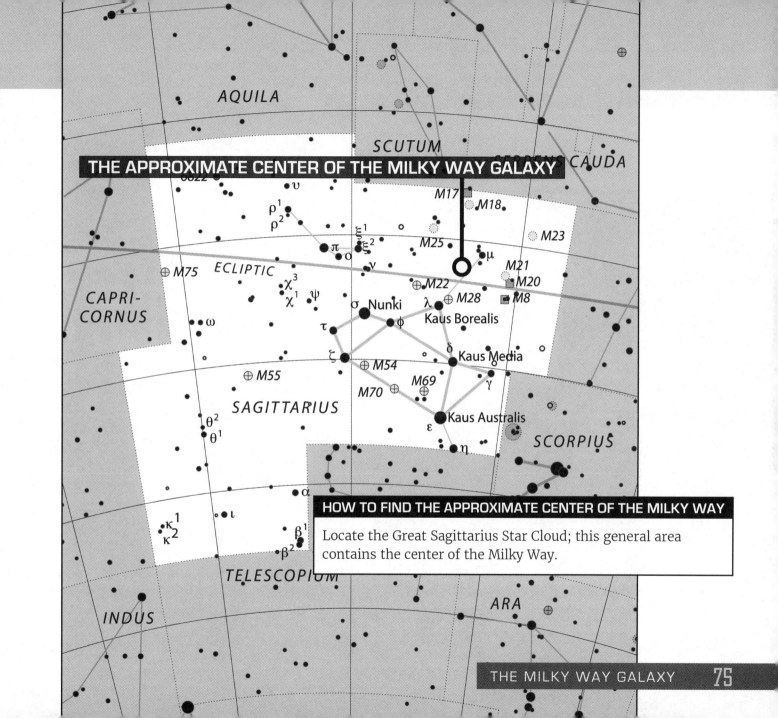

THE APPROXIMATE CENTER OF THE MILKY WAY GALAXY

AQUILA

SCUTUM

CAUDA

NGC6822

υ

ρ¹
ρ²

ξ¹
ξ²

π

ο

ν

M17

M18

M23

M25

μ

M21

M20

M22

M28

M8

M75

ECLIPTIC

CAPRI-
CORNUS

χ³
χ¹ ψ

ω

σ Nunki

φ

τ

ζ

M54

M55

θ²
θ¹

M70

M69

SAGITTARIUS

α

κ¹
κ²

ι

β¹

β²

TELESCOPIUM

INDUS

λ

Kaus Borealis

δ Kaus Media

γ

Kaus Australis

ε

η

SCORPIUS

ARA

HOW TO FIND THE APPROXIMATE CENTER OF THE MILKY WAY

Locate the Great Sagittarius Star Cloud; this general area contains the center of the Milky Way.

CHAPTER SEVEN

STAR CLUSTERS

So far, you have looked at very large and messy formations in space. Now you are going to focus in on smaller, more neatly defined formations. And there is an irony in the fact that many of these neat and compact objects are actually called Messier objects!

This isn't because of their appearance—it's because of the man, Charles Messier, who discovered and cataloged them in the late-eighteenth century.

Before the invention of the telescope, many of these objects were thought to be stars. Now that we can view them through a telescope, we have discovered that many of these "stars" are actually very complex, very large structures. Indeed, the Milky Way galaxy is a growing, evolving, and changing place with a variety of interesting objects. Star clusters are a good example of this.

TYPES OF STAR CLUSTERS

Star clusters are categorized into several different types, and you will be examining two with your telescope: 1) globular clusters, which are tightly packed formations of stars, and 2) open clusters, which are loosely formed groups of stars. Finally, you will observe a rare double cluster, which is a pair of open clusters that are very close together.

GLOBULAR STAR CLUSTERS

A globular star cluster is a spherical formation of stars; it might look like a symmetrical cotton ball. These formations are tightly bound by gravity, with bright, dense centers and less dense outer areas.

Typically, they are composed of tens of thousands, occasionally millions, of stars. For most of them, you will be able to make out their shape with a small telescope. And there are a few that you can zoom in on to reveal separations of various stars. Either way, a globular star cluster is an amazing sight to see.

OPEN STAR CLUSTERS

These are loosely arranged formations of stars, typically with fewer than one hundred stars. There are two different types of open star clusters. The first type, which is more common, is a system of stars that are loosely bound together by gravitational attraction. The second type is an optical cluster. The stars are not near each other in space; they just appear to be near each other because of our perspective here on Earth.

Open clusters might not appear to be as magnificent as globular clusters but they are much larger in size and much easier to find. And, once found, your skill with the telescope and star charts will have improved. You will be able to go back to the globular clusters and find any that you previously couldn't find.

21. M13

OBJECT TYPE: Globular star cluster

CONSTELLATION: Hercules

APPARENT MAGNITUDE: +5.8

COORDINATES: 16h 41m 24 s, +36° 27' 35.5"

SEASON: April to November; best viewing in summer

DIFFICULTY: Easy

Many amateur astronomers consider this to be the best globular cluster in the northern sky because it is very easy to find, very bright, and very uniform in its globular shape. On a good viewing night, with a clear and dark sky, it can be seen faintly with the naked eye. It is composed of half a million stars and it is very symmetrical in appearance. With a small telescope, you probably cannot resolve it to individual stars. It will look like a cotton ball. But, it is well worth finding, and reasonably easy to locate right in the armpit of Hercules.

With your telescope can you make out some individual stars? If you can, then that is great. You have got a nice little telescope there!

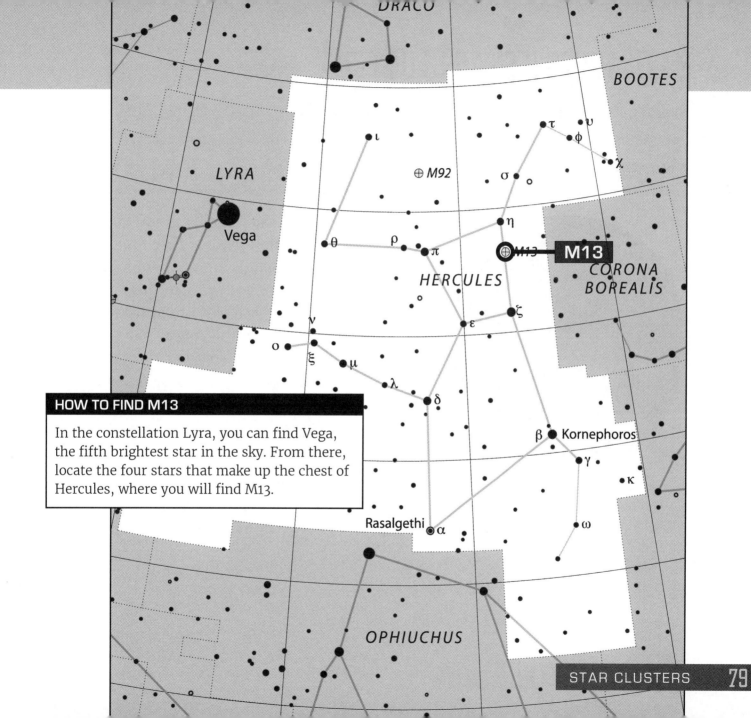

HOW TO FIND M13

In the constellation Lyra, you can find Vega, the fifth brightest star in the sky. From there, locate the four stars that make up the chest of Hercules, where you will find M13.

22. M4 GLOBULAR STAR CLUSTER

OBJECT TYPE: Globular star cluster

CONSTELLATION: Scorpius

APPARENT MAGNITUDE: +5.9

COORDINATES: 16h 23m 35.22s, -26° 31' 32.7"

SEASON: June through August; best in July

DIFFICULTY: Medium

This cluster, which is only about 7,000 light years away, is the closest to us here on Earth. Scorpius is a constellation in the southern sky, so it can be limited in view if you live in the northern part of the United States. Much of it can be below the horizon. But, if within viewing range, M4 is an excellent cluster to find if you are a beginner. This is because of its proximity to the very bright and very red star Antares.

With this act of using a telescope, you have spanned centuries. What people in civilizations past had seen was a star. But, when you turn your telescope to M4, it is revealed to be a complex cluster of stars.

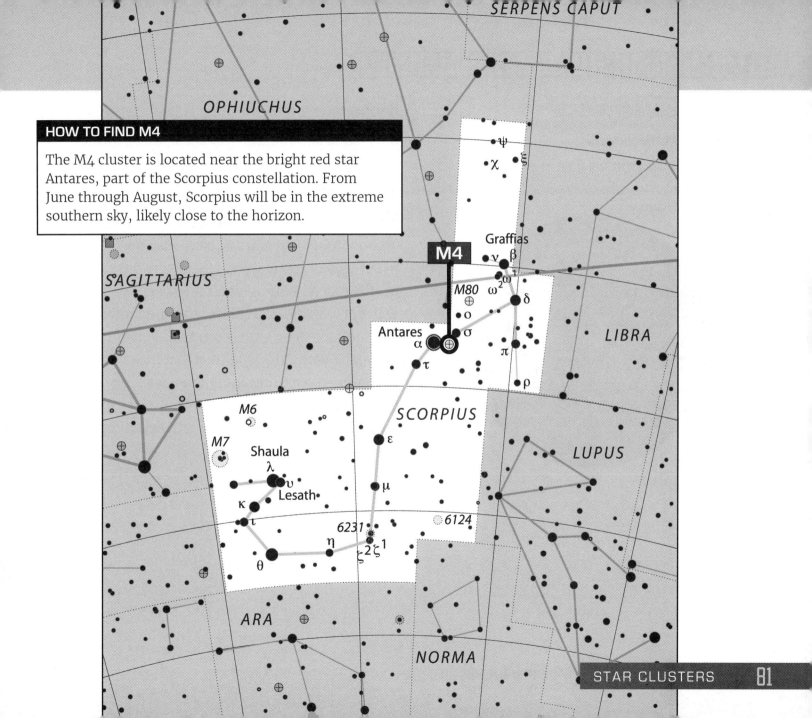

SERPENS CAPUT

OPHIUCHUS

HOW TO FIND M4

The M4 cluster is located near the bright red star
Antares, part of the Scorpius constellation. From
June through August, Scorpius will be in the extreme
southern sky, likely close to the horizon.

ψ
χ   ξ

Graffias
ν β
M4
ω² ω¹ τ
M80
δ

SAGITTARIUS

o

Antares   σ
α   π
τ   ρ

LIBRA

SCORPIUS

M6

ε

M7

Shaula
λ
υ
Lesath
κ   μ   6124
ι
6231
η
ζ² ζ¹
θ

LUPUS

ARA

NORMA

23. M15 GLOBULAR STAR CLUSTER

OBJECT TYPE: Globular star cluster

CONSTELLATION: Pegasus

APPARENT MAGNITUDE: +6.2

COORDINATES: 21h 29m 58.33s, +12° 10' 01.2"

SEASON: August through November; best in October

DIFFICULTY: Easy

This cluster of stars is estimated to contain about 30,000 individual stars. It is the densest star cluster in our galaxy and is second in density only to the core of the galaxy. With a small telescope, you probably will not be able to resolve any individual stars. It will look like a bright cotton ball. But, if you have a telescope with an aperture of 5 inches or more, you will start to make out individual stars. Don't let this density intimidate you: With any telescope or even a pair of binoculars, this is an excellent celestial object to observe.

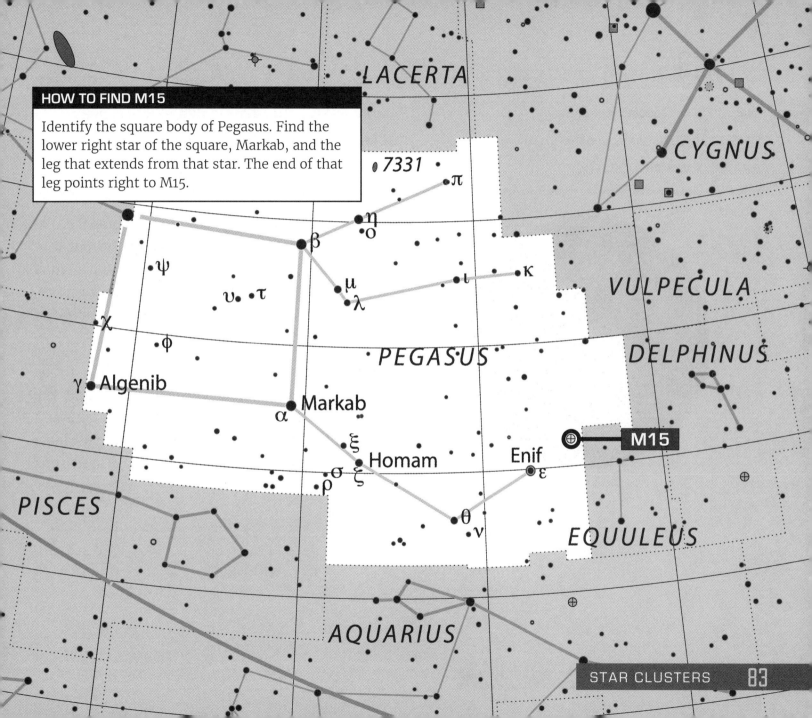

HOW TO FIND M15

Identify the square body of Pegasus. Find the lower right star of the square, Markab, and the leg that extends from that star. The end of that leg points right to M15.

LACERTA

CYGNUS

7331

π

η

β

ο

VULPECULA

ψ

μ

ι

κ

υ

τ

λ

χ

DELPHINUS

φ

PEGASUS

γ

Algenib

Markab

α

ξ

Homam

Enif

ρ

σ

ξ

ε

PISCES

θ

ν

EQUULEUS

M15

AQUARIUS

24. M3

OBJECT TYPE: Globular star cluster

CONSTELLATION: Canes Venatici

APPARENT MAGNITUDE: +6.2

COORDINATES: 13h 42m 11.62s, +28° 22' 38.2"

SEASON: Available year-round for most of North America

DIFFICULTY: Easy

This is a remarkable star cluster composed of more than 200,000 stars; some estimates put it at over 500,000 stars. It is a great target for a small telescope, and with a telescope of 6 inches or more, you should be able to resolve many of the outer individual stars.

Amateur astronomers often consider this to be the second best globular cluster in the night sky because of the large number of stars it contains and its overall brightness. Plus, this cluster is available year-round, which makes it a great choice in seasons other than summer.

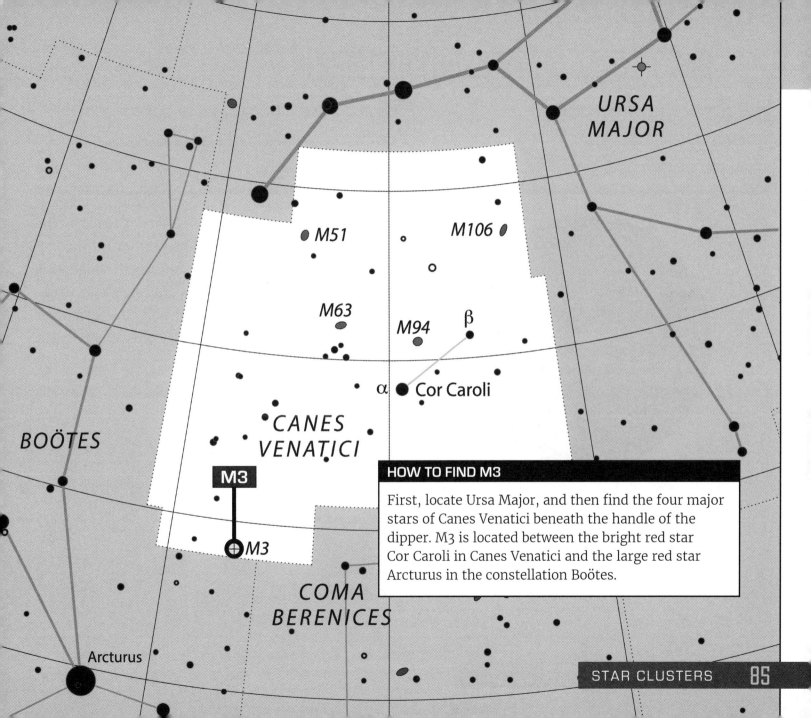

URSA MAJOR

M51

M106

M63

M94

β

α Cor Caroli

CANES VENATICI

BOÖTES

M3

M3

COMA BERENICES

Arcturus

HOW TO FIND M3

First, locate Ursa Major, and then find the four major stars of Canes Venatici beneath the handle of the dipper. M3 is located between the bright red star Cor Caroli in Canes Venatici and the large red star Arcturus in the constellation Boötes.

25. ELLIPTICAL STAR CLUSTER (M22)

OBJECT TYPE: Elliptical star cluster

CONSTELLATION: Sagittarius

APPARENT MAGNITUDE: +5.1

COORDINATES: 18h 36m 23.94s, -23° 54' 17.1"

SEASON: June through early September

DIFFICULTY: Easy

This cluster is commonly referred to as the Sagittarius cluster because of its location. Almost all globular star clusters are very uniform in shape; this one's elliptical shape is unique. Having an apparent magnitude of 5.1 makes it one of the brightest clusters in the sky and reasonably easy to spot with the naked eye.

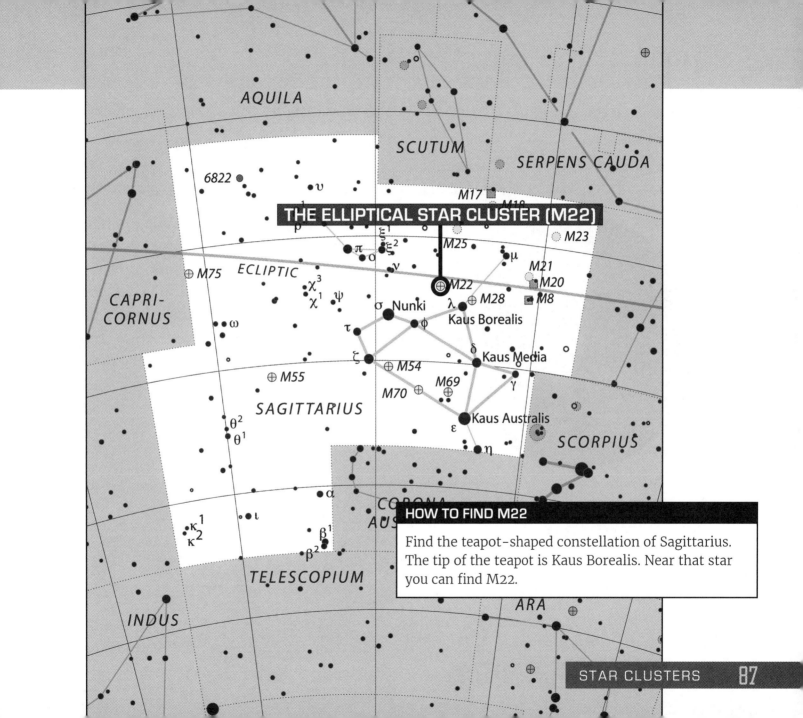

THE ELLIPTICAL STAR CLUSTER (M22)

AQUILA

SCUTUM

SERPENS CAUDA

6822

υ

ρ

M17

M18

M23

M25

ξ¹
ξ²

π
o
ν

μ

M21

M20

ECLIPTIC

M75

χ³
χ¹
ψ

M22

M28

M8

CAPRI-
CORNUS

σ Nunki

λ

Kaus Borealis

ω

τ

δ Kaus Media

ζ

M54

γ

M55

M70

M69

SAGITTARIUS

θ²
θ¹

Kaus Australis

ε

SCORPIUS

η

κ¹
κ²

β¹

CORONA
AUS

β²

ι

α

TELESCOPIUM

HOW TO FIND M22

Find the teapot-shaped constellation of Sagittarius.
The tip of the teapot is Kaus Borealis. Near that star
you can find M22.

INDUS

ARA

26. M92

OBJECT TYPE: Globular star cluster

CONSTELLATION: Hercules

APPARENT MAGNITUDE: +6.3

COORDINATES: 17h 17m 07.39s, +43° 08' 09.4"

SEASON: April to November; best in summer

DIFFICULTY: Easy

M92 is worthy of note because is one of the brightest globular clusters in the night sky. However, it's often overlooked because of its proximity to the impressive M13. It is magnitude 6.3 and M13 is magnitude 5.8, so M92 is just slightly dimmer in appearance. It is estimated to be over 14 billion years old, which would make it the oldest globular cluster in the Milky Way.

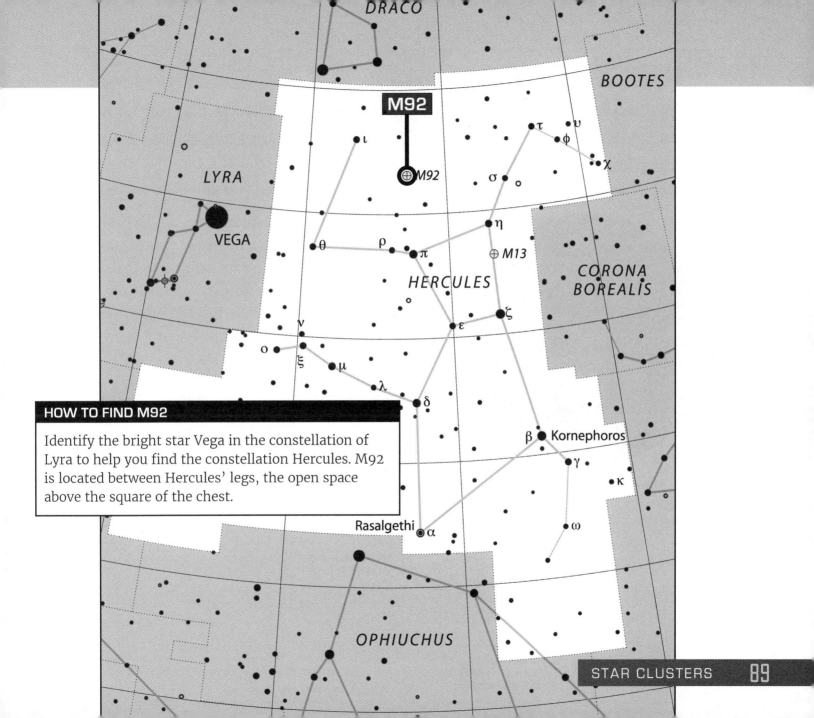

M92

M92

DRACO

BOOTES

τ υ
φ
χ

ι

LYRA

σ ο

η

VEGA

θ ρ
π

⊕ M13

HERCULES

CORONA
BOREALIS

ν

ζ
ε

ο
ξ

μ

λ

δ

β Kornephoros

γ
κ

HOW TO FIND M92

Identify the bright star Vega in the constellation of
Lyra to help you find the constellation Hercules. M92
is located between Hercules' legs, the open space
above the square of the chest.

ω

Rasalgethi α

OPHIUCHUS

27. M5

OBJECT TYPE: Globular star cluster

CONSTELLATION: Serpens Caput

APPARENT MAGNITUDE: +6.65

COORDINATES: 15h 18m 33.22s, +02° 04' 51.7"

SEASON: June through September; best in July

DIFFICULTY: Medium

Approximately 13 billion years old, M5 is just at the range of visibility to the naked eye on good viewing nights, so you may have to find it first with your finder scope or a pair of binoculars. Use your lowest power eyepiece to first locate it then move on to higher power eyepieces to try and resolve it into separate stars.

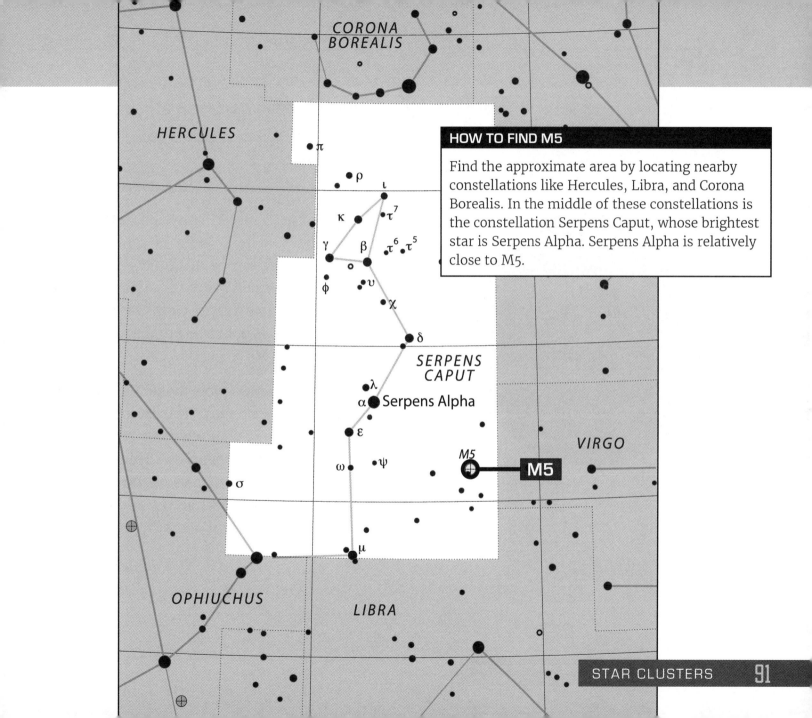

CORONA BOREALIS

HERCULES

π

ρ
ι
κ
τ⁷
γ   β
τ⁶ τ⁵
ο
φ
υ
χ
δ

SERPENS CAPUT

λ
α● Serpens Alpha
ε
ω ψ
σ
M5
μ

VIRGO

OPHIUCHUS

LIBRA

HOW TO FIND M5

Find the approximate area by locating nearby constellations like Hercules, Libra, and Corona Borealis. In the middle of these constellations is the constellation Serpens Caput, whose brightest star is Serpens Alpha. Serpens Alpha is relatively close to M5.

M5

28. M80

OBJECT TYPE: Globular star cluster

CONSTELLATION: Scorpius

APPARENT MAGNITUDE: +7.87

COORDINATES: 16h 17m 02.41s, -22° 58' 33.9"

SEASON: June through August; best in July

DIFFICULTY: Medium

This cluster has a nice density, but with a small telescope, it will only be seen as a fuzzy cluster. It takes a larger telescope of 12 inches or more to resolve some of its individual stars. It is nonetheless an excellent star cluster to view. Because it is a magnitude 7.87, it is not visible to the naked eye, so it may be challenging to locate.

If you have found M80, you should be very proud. This is a real astronomy task to accomplish. You actually have seen, in your telescope, an object that cannot be seen with the naked eye. Think about that for just a moment. No human being, throughout the history of mankind has ever seen M80, until recent times, thanks to the invention of the telescope.

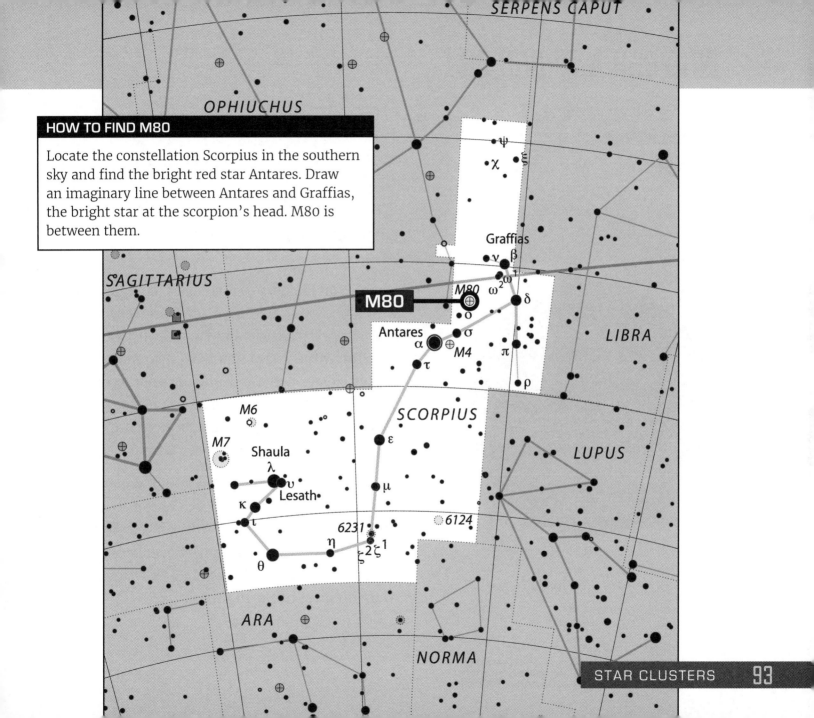

HOW TO FIND M80

Locate the constellation Scorpius in the southern sky and find the bright red star Antares. Draw an imaginary line between Antares and Graffias, the bright star at the scorpion's head. M80 is between them.

OPHIUCHUS

SERPENS CAPUT

ψ
χ ξ

Graffias
ν β

ω² ω¹
M80 ω τ
δ
M80

SAGITTARIUS

ο

Antares σ
α M4
τ π

LIBRA

ρ

SCORPIUS

M6

M7
Shaula
λ
υ
κ Lesath
ι
τ

ε

μ

6124

6231
η
ζ2 ζ1
θ

LUPUS

ARA

NORMA

29. NGC 869 AND NGC 884

OBJECT TYPE: Double star cluster

CONSTELLATION: Perseus

APPARENT MAGNITUDE OF NGC 869: +3.7

APPARENT MAGNITUDE OF NGC 884: +3.8

COORDINATES OF NGC 869: 02h 19.1m 00s, +57° 09' 00"

COORDINATES OF NGC 884: 02h 22.0m 00s, +57° 08' 00"

SEASON: Available year-round for most of North America

DIFFICULTY: Easy

Double star clusters that are close enough to appear within the same telescope view are very rare. This one is a truly excellent sight and a must-see because of its brightness and unique appearance. View this double cluster with binoculars or your lowest power eyepiece first to get a sense for it. Then you can change to a higher power eyepiece to explore the complexity and differences of each.

If you are an absolute beginner with your telescope, this is an excellent deep space object to start with because of its brightness. You can easily find this double cluster with your naked eye. Because it is in the constellation of Perseus, it is available year-round to most of North America.

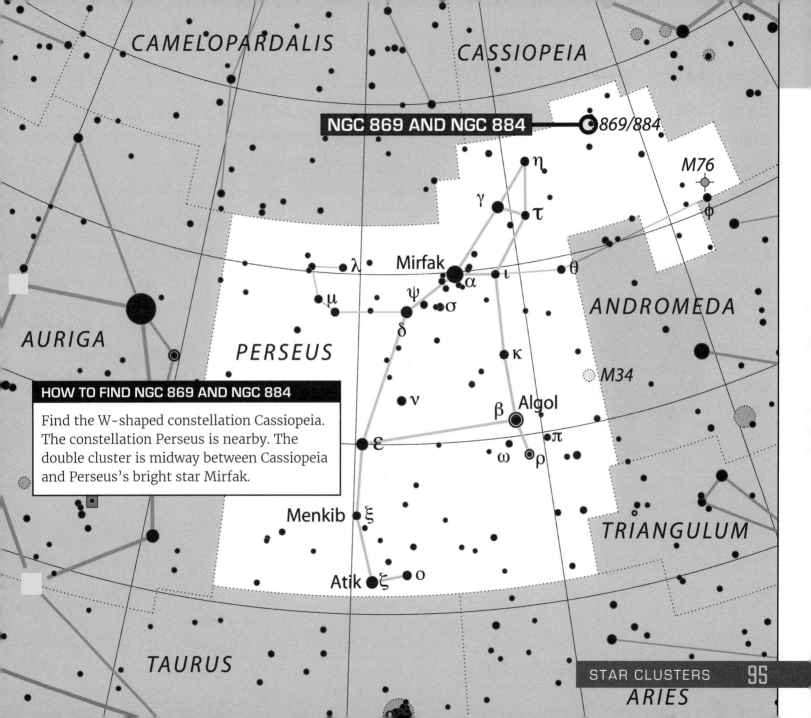

CAMELOPARDALIS

CASSIOPEIA

NGC 869 AND NGC 884 ─── ◎ *869/884*

M76

η

γ τ

φ

λ Mirfak

μ ψ α ι θ

σ

δ

ANDROMEDA

AURIGA

κ

PERSEUS

M34

ν

β Algol

HOW TO FIND NGC 869 AND NGC 884

Find the W-shaped constellation Cassiopeia.
The constellation Perseus is nearby. The
double cluster is midway between Cassiopeia
and Perseus's bright star Mirfak.

π

ω ρ

ε

Menkib ● ξ

TRIANGULUM

Atik ● ζ ● ο

TAURUS

ARIES

30. THE PLEIADES (M45)

OBJECT TYPE: Open star cluster

CONSTELLATION: Taurus

APPARENT MAGNITUDE: +1.6

COORDINATES: 3h 47m 24s, +24° 07' 00"

SEASON: November to March; best in January

DIFFICULTY: Easy

This is the most famous of all the open clusters, often referred to as "the Seven Sisters."

This asterism of stars is an excellent test of your eyesight and of the viewing conditions where you live. Typically, to the naked eye, you will only see five stars, and on good viewing nights you may see six. If you can see seven stars, then you have great vision and terrific dark sky viewing conditions.

You don't need to find this open cluster with your telescope. It is very large and you can easily find it first with just your naked eyes. Once you find it, turn your telescope up and adventure around, exploring its complexity and beauty.

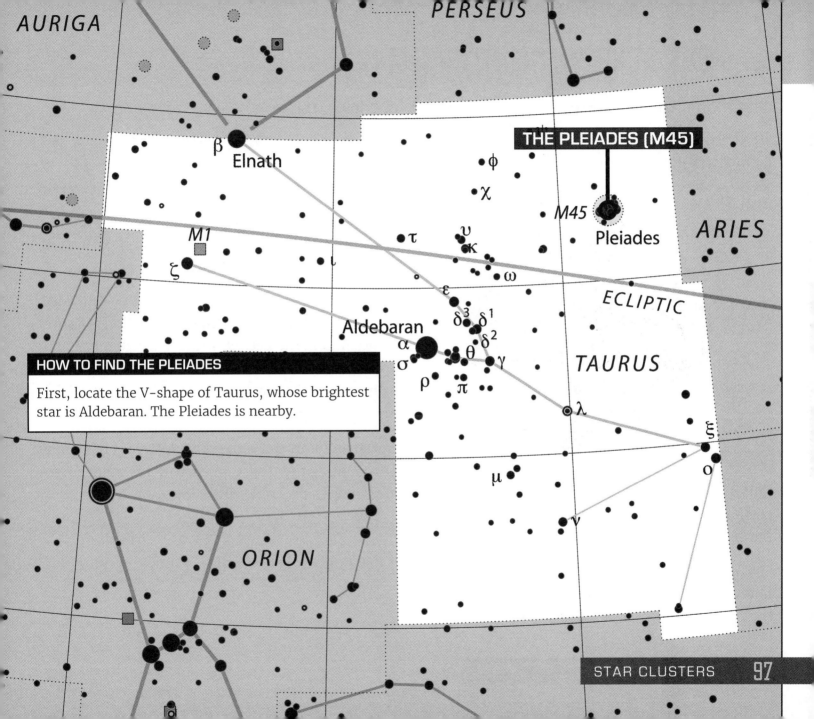

AURIGA

PERSEUS

β
Elnath

THE PLEIADES (M45)

φ

χ

M45

Pleiades

ARIES

M1

ζ

ι

τ

υ
κ

ω

ECLIPTIC

ε

δ³ δ¹

Aldebaran

δ²

α

θ

γ

TAURUS

σ

ρ

π

λ

HOW TO FIND THE PLEIADES

First, locate the V-shape of Taurus, whose brightest
star is Aldebaran. The Pleiades is nearby.

ξ

o

μ

γ

ORION

31. THE BEEHIVE CLUSTER (M44)

OBJECT TYPE: Open star cluster

CONSTELLATION: Cancer

APPARENT MAGNITUDE: +3.7

COORDINATES: 08h 40.4m 00s, 19° 59' 00"

SEASON: Late January through early June

DIFFICULTY: Easy

This is a very loose, very bright, and very open cluster. It spans a width of sky almost three times the size of the moon.

For many objects in this book, the larger your telescope, the better the viewing. But, that is not the case with this cluster. It is an excellent object for viewing with small telescopes and low powers because it is bright and dispersed. It is also an excellent choice for binoculars.

Don't overlook the Beehive Cluster. It is well worth finding. And even though the constellation of Cancer is rather dim, the Beehive Cluster is rather bright. It is quite possible that you'll find the cluster first and from there, you'll be able to identify the stars that make up Cancer.

How many stars can you resolve with your telescope? Try counting them up. Galileo, with his small homemade telescope, could count forty.

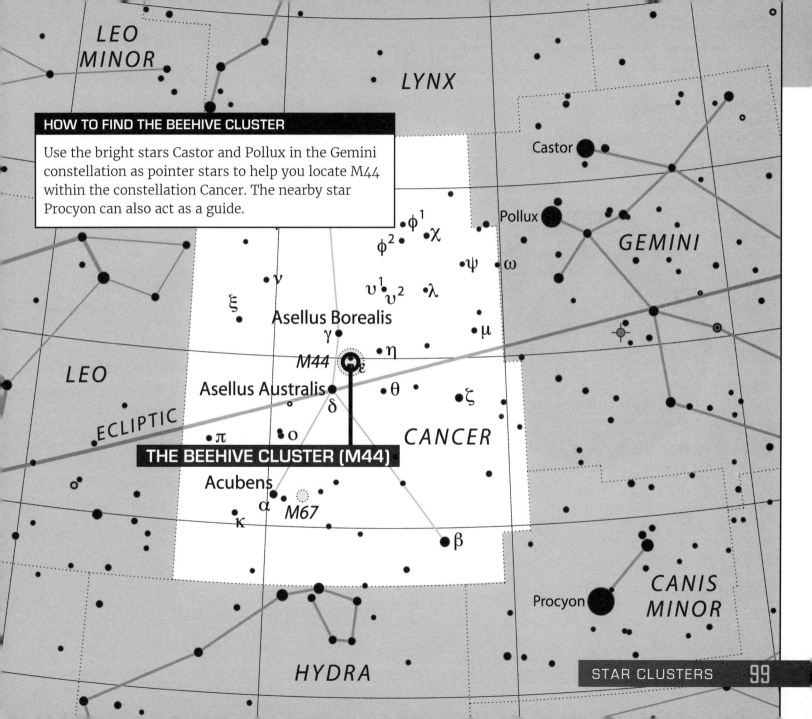

HOW TO FIND THE BEEHIVE CLUSTER

Use the bright stars Castor and Pollux in the Gemini constellation as pointer stars to help you locate M44 within the constellation Cancer. The nearby star Procyon can also act as a guide.

LEO MINOR

LYNX

LEO

Castor

Pollux

GEMINI

φ¹ χ

φ²

ψ ω

ν

υ¹ υ² λ

ξ

μ

Asellus Borealis

γ η

M44 ε

Asellus Australis

θ ζ

δ

ECLIPTIC

CANCER

π ο

THE BEEHIVE CLUSTER (M44)

Acubens

α M67

κ

β

Procyon

CANIS MINOR

HYDRA

32. THE PTOLEMY CLUSTER [M7]

OBJECT TYPE: Open star cluster

CONSTELLATION: Scorpius

APPARENT MAGNITUDE: +3.3

COORDINATES: 17h 53m 51.20s, -34° 47' 34"

SEASON: July through August

DIFFICULTY: Easy

The Ptolemy Cluster is composed of a total of about eighty stars, not all of which are visible with a small telescope. This is a great way to test your telescope and to get a sense for the viewing conditions. The more stars you see, the better your telescope and viewing conditions.

This cluster is in the southernmost part of Scorpius, which means it will be low in the sky and potentially near the horizon. So, it is visible only at certain times, depending on how far north in North America you are located. The further south you are, the longer it is available during the year.

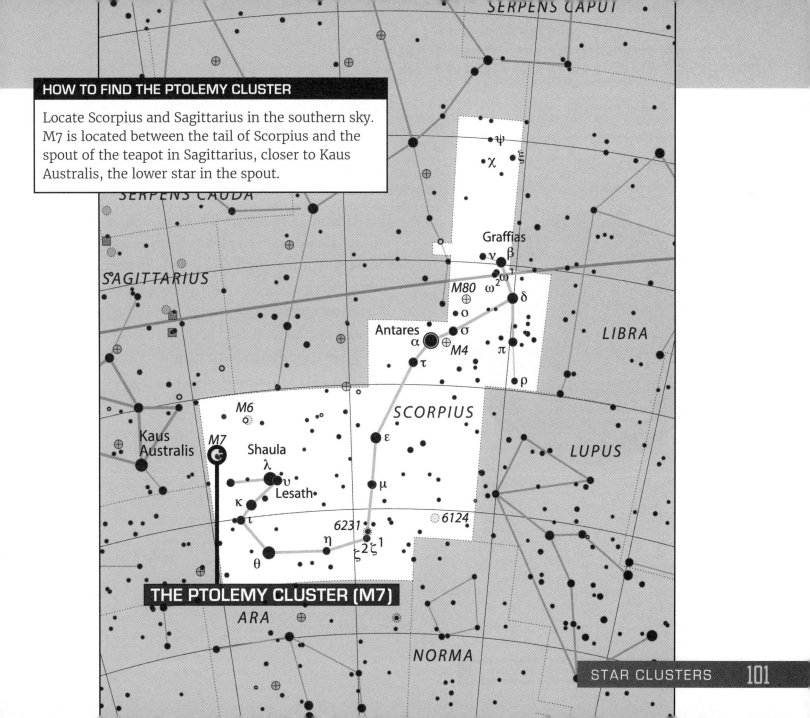

HOW TO FIND THE PTOLEMY CLUSTER

Locate Scorpius and Sagittarius in the southern sky. M7 is located between the tail of Scorpius and the spout of the teapot in Sagittarius, closer to Kaus Australis, the lower star in the spout.

33. THE BUTTERFLY CLUSTER (M6)

OBJECT TYPE: Open star cluster

CONSTELLATION: Scorpius

APPARENT MAGNITUDE: +4.2

COORDINATES: 17h 40.1m 00s, -32° 13' 00"

SEASON: July through August

DIFFICULTY: Easy

This cluster is made up of about eighty stars and is best viewed with the lowest power eyepiece that you have. The cluster is dispersed, and using a lower power will keep it all in view. It is called the Butterfly Cluster because it has a rough, two-lobed shape like that of a butterfly. It is in southern Scorpius, so you only have a limited viewing time if you are in the northern part of the United States.

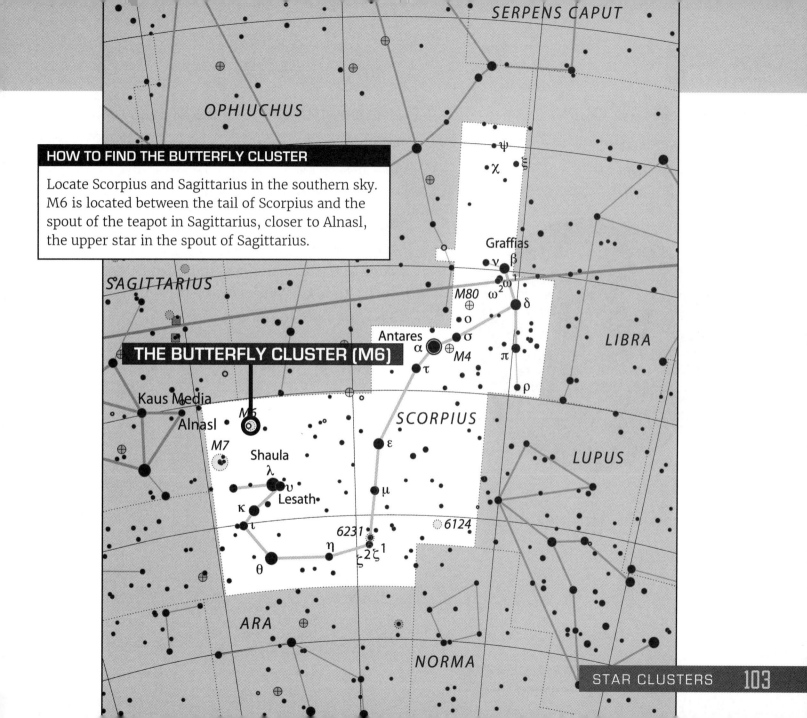

HOW TO FIND THE BUTTERFLY CLUSTER

Locate Scorpius and Sagittarius in the southern sky. M6 is located between the tail of Scorpius and the spout of the teapot in Sagittarius, closer to Alnasl, the upper star in the spout of Sagittarius.

THE BUTTERFLY CLUSTER (M6)

34. M52

OBJECT TYPE: Open star cluster

CONSTELLATION: Cassiopeia

APPARENT MAGNITUDE: +5.0

COORDINATES: 23h 24.2m 00s, +61° 35' 00"

SEASON: Available year-round for most of North America

DIFFICULTY: Easy

This is considered to be a very fine cluster to observe, and it is also easy to locate. It is a definite must if you want to observe an open star cluster that isn't too open. It almost verges on being a globular cluster, but isn't quite there.

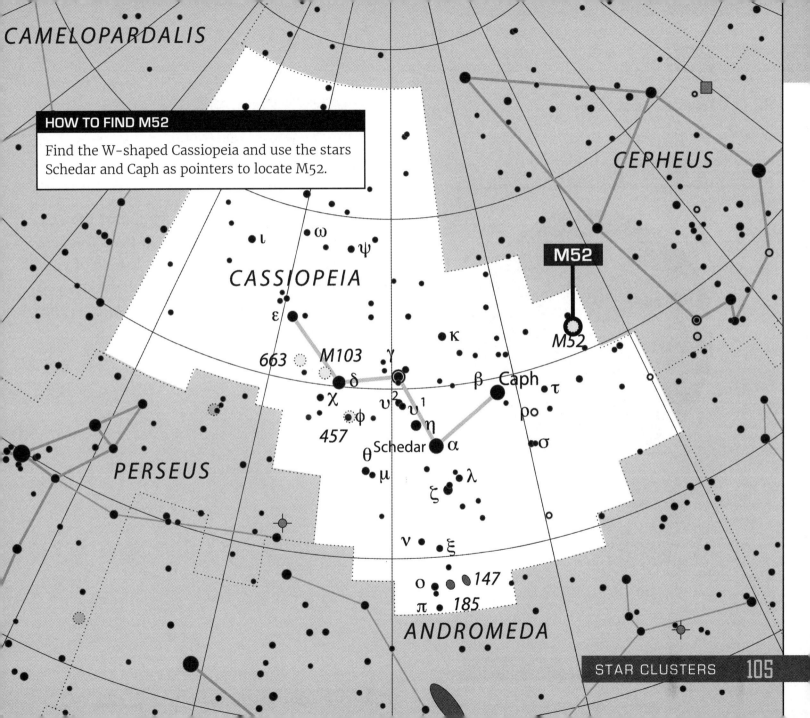

CAMELOPARDALIS

CASSIOPEIA

CEPHEUS

PERSEUS

ANDROMEDA

HOW TO FIND M52

Find the W-shaped Cassiopeia and use the stars Schedar and Caph as pointers to locate M52.

M52

M52

ι ω ψ

ε

663 M103 γ

δ

κ

χ υ² υ¹ β Caph τ

457 φ η ρ

θ Schedar α σ

μ λ

ζ

ν ξ

ο 147

π 185

35. M35

OBJECT TYPE: Open star cluster

CONSTELLATION: Gemini

APPARENT MAGNITUDE: +5.3

COORDINATES: 06h 09.1m 00s, +24° 21' 00"

SEASON: November through April

DIFFICULTY: Easy

M35 is scattered in an area of sky almost the size of the moon, which means you will need to use a low-power eyepiece to see it all in one field of view without moving the telescope. Under dark skies you should be able to find it with the naked eye. This makes it an easy target.

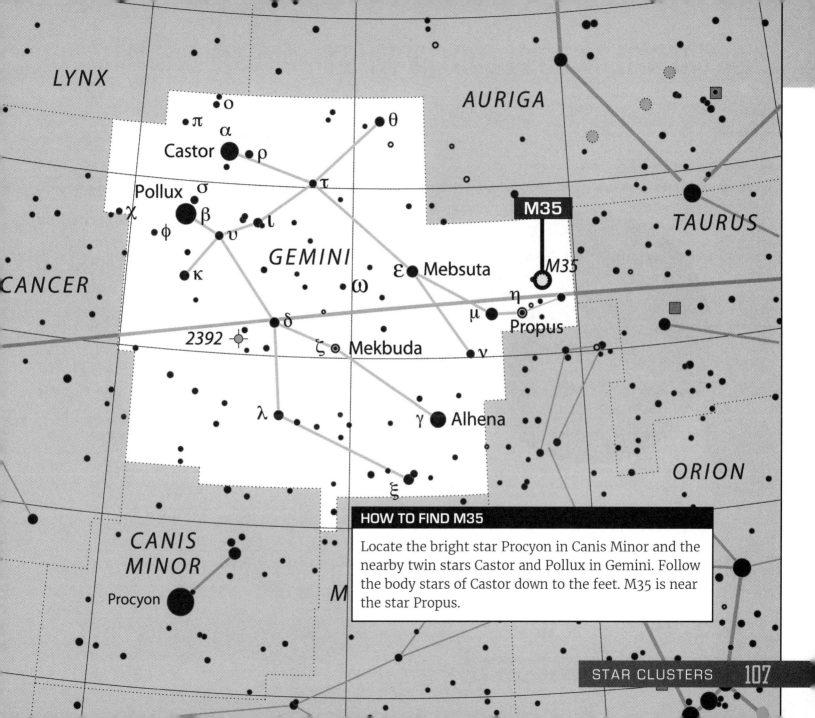

LYNX

AURIGA

ο
π
α
Castor
ρ
θ

Pollux σ
χ
β
φ ι υ
κ

CANCER

GEMINI

ω
ε Mebsuta

M35

M35

η
μ
Propus

2392
δ
ζ Mekbuda
γ

λ
γ Alhena

TAURUS

ORION

ξ

**CANIS
MINOR**

Procyon

HOW TO FIND M35

Locate the bright star Procyon in Canis Minor and the
nearby twin stars Castor and Pollux in Gemini. Follow
the body stars of Castor down to the feet. M35 is near
the star Propus.

36. THE WILD DUCK CLUSTER [M11]

OBJECT TYPE: Open star cluster

CONSTELLATION: Scutum

APPARENT MAGNITUDE: +6.3

COORDINATES: 18h 51.1m 00s, -06° 16' 00"

SEASON: August through October; best in August

DIFFICULTY: Easy

This is a definite for small telescopes. M11 is a dense open cluster with about 3,000 stars. It is one of the richest and the densest of all the open clusters and comes the closest to being a globular cluster.

It is called the Wild Duck Cluster because with a small telescope, the brightest stars form the shape of a V, just like a flock of flying ducks would. You can get a bit of a sense for this with this circular image of it. But, with this image, many more of the dimmer stars can also be seen.

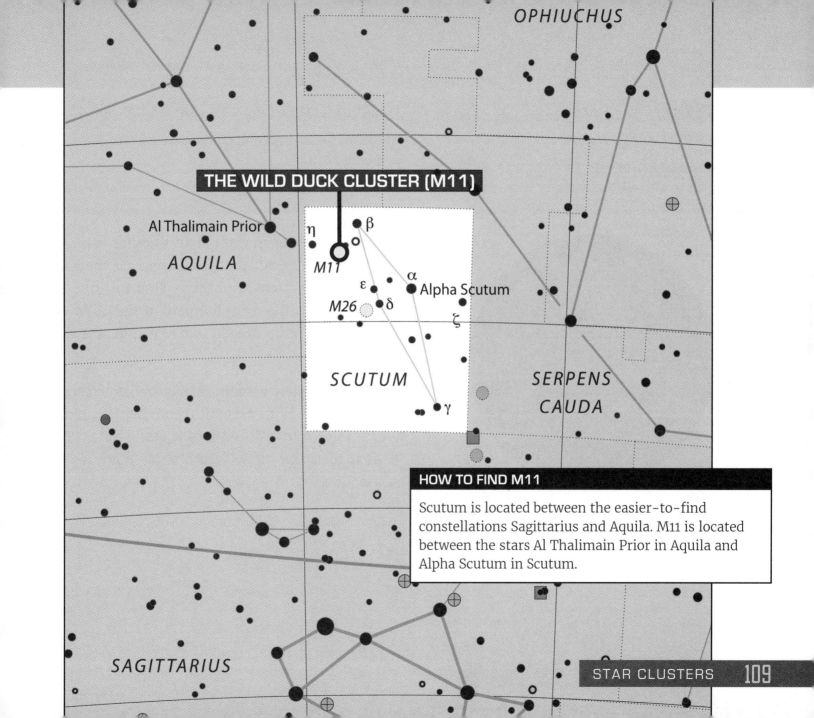

OPHIUCHUS

THE WILD DUCK CLUSTER (M11)

Al Thalimain Prior

η

β

○

M11

AQUILA

α

Alpha Scutum

ε

M26 ○ δ

ζ

SCUTUM

γ

SERPENS
CAUDA

HOW TO FIND M11

Scutum is located between the easier-to-find
constellations Sagittarius and Aquila. M11 is located
between the stars Al Thalimain Prior in Aquila and
Alpha Scutum in Scutum.

SAGITTARIUS

37. M37

OBJECT TYPE: Open star cluster

CONSTELLATION: Auriga

APPARENT MAGNITUDE: +6.2

COORDINATES: 5h 52m 18s, +32° 33' 02"

SEASON: December through April; best in late February and early March

DIFFICULTY: Easy

There are three open star clusters not far from each other in the constellation of Auriga. Of the three, M37 is the best one to observe. Find this one first and you are in the neighborhood of the other two. It is well worth the time to find M36 and M38 while you are there.

This is an excellent example of how bigger telescopes can see more. With a small telescope, this cluster will resolve into a dozen bright stars. But with an 8-inch telescope, it will resolve as many as 200 stars.

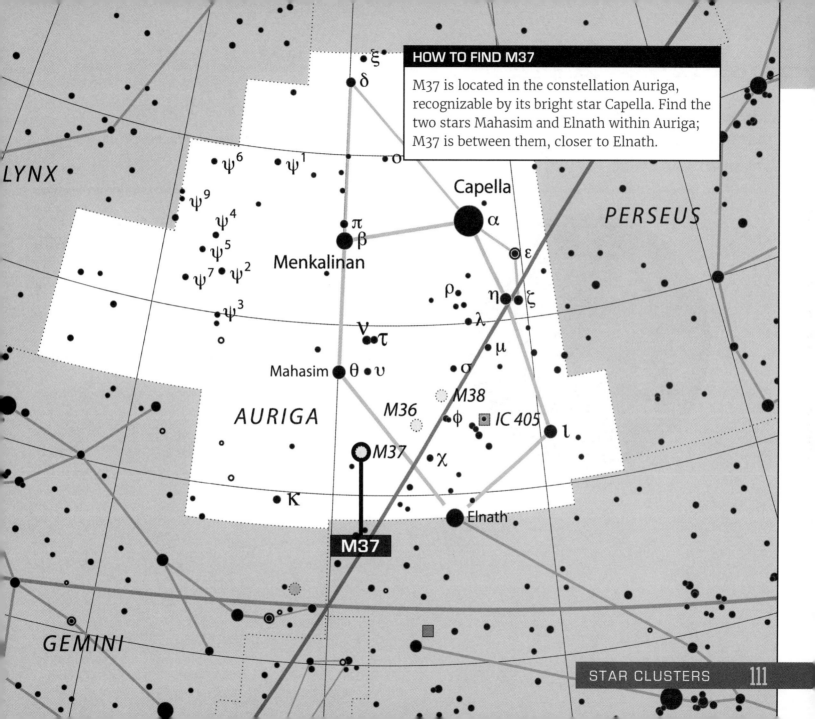

LYNX

ξ
δ

HOW TO FIND M37

M37 is located in the constellation Auriga, recognizable by its bright star Capella. Find the two stars Mahasim and Elnath within Auriga; M37 is between them, closer to Elnath.

ο

ψ⁶   ψ¹

ψ⁹

ψ⁴

Capella

PERSEUS

π
β
Menkalinan

α

ε

ψ⁵

ψ⁷   ψ²

ρ

η   ξ
λ

ψ³

ν
τ

μ

Mahasim   θ   υ

σ

AURIGA

M38

M36   φ   IC 405

ι

M37

χ

κ

Elnath

M37

GEMINI

38. M41

OBJECT TYPE: Open star cluster

CONSTELLATION: Canis Major

APPARENT MAGNITUDE: +4.6

COORDINATES: 06h 46.0m, -20° 46' 00"

SEASON: November through March; best in January

DIFFICULTY: Easy

This open cluster is a good viewing choice. It is easy to see and find because it is just south of Sirius. With a small telescope and good viewing conditions, you should be able to resolve this cluster into about fifty stars. With a medium telescope, you will be able to discern about 100 stars. This open star cluster is also a good opportunity for you to practice the technique of averted vision.

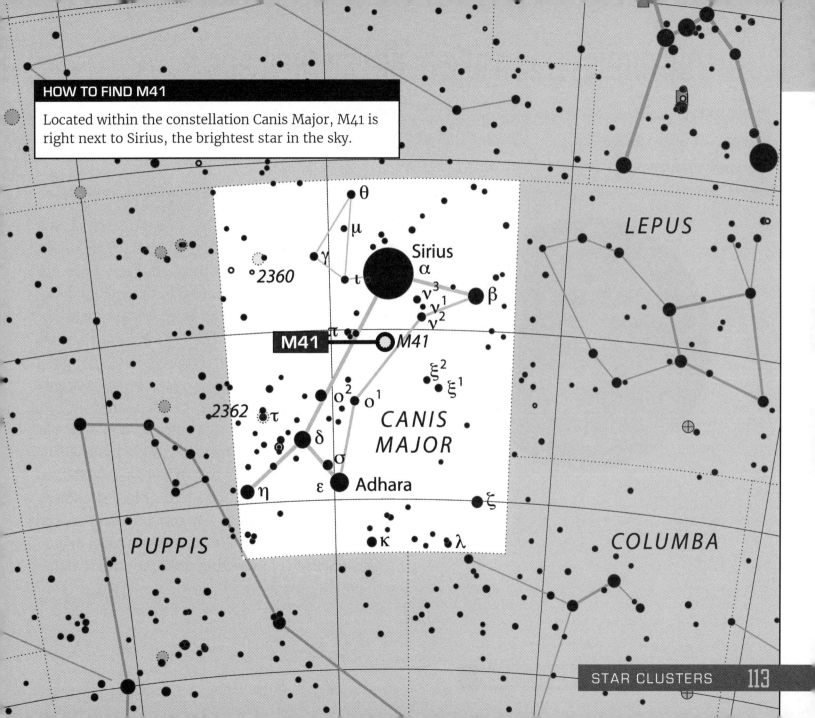

HOW TO FIND M41

Located within the constellation Canis Major, M41 is right next to Sirius, the brightest star in the sky.

LEPUS

θ

μ

γ

ι

Sirius

α

2360

ν³
ν¹
ν²

β

π

M41

M41

ξ²
ξ¹

2362 τ

o² o¹

CANIS
MAJOR

ϱ δ

σ

η ε Adhara

ζ

κ λ

PUPPIS

COLUMBA

39. THE CHRISTMAS TREE CLUSTER (NGC 2264)

OBJECT TYPE: Open star cluster with nebulosity

CONSTELLATION: Monoceros

APPARENT MAGNITUDE: +3.9

COORDINATES: 06h 40m 58s, +09 °53' 42"

SEASON: November through March; best in February

DIFFICULTY: Easy

The Christmas Tree Cluster is a rare set of star clusters that is embedded in a nebula. This makes for an interesting view through your telescope because the stars add light to the surrounding gases of the nebula, illuminating it. The stars make the nebula glow, giving you a better look. With a small telescope and low-power eyepiece, the triangular Christmas-tree shape can easily be seen. It is similar to the M42 Great Nebula in Orion because it, too, consists of a cluster of stars embedded in a nebula. The bright star in the image is 15 Monocerotis, and it marks the trunk of the tree. It is a blue-white star of magnitude 4.6. Note that for this night sky object, orientation is important. For some telescopes, you will see it upside-down. If it is upside-down in your telescope, use your erecting prism to turn it right-side up!

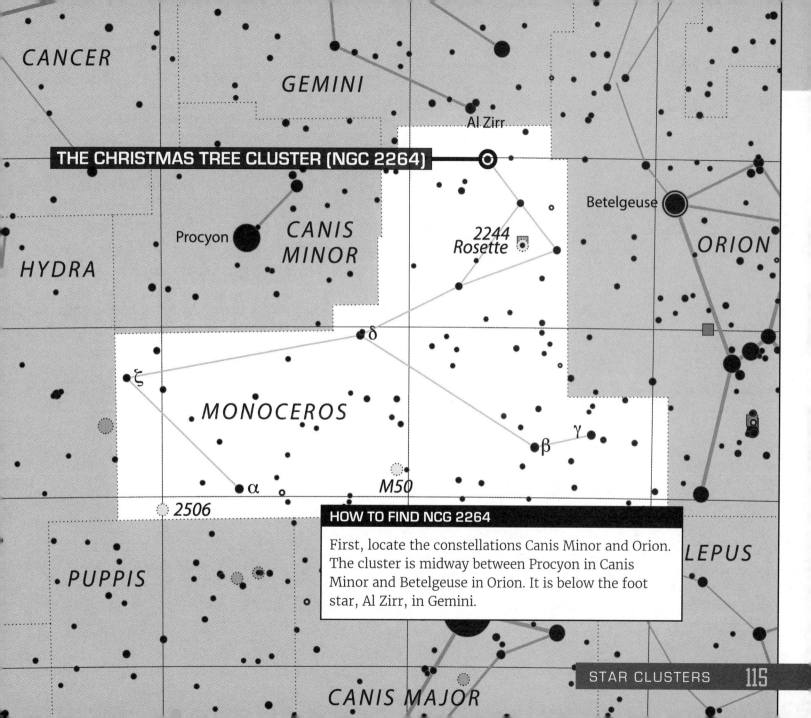

CANCER

GEMINI

Al Zirr

THE CHRISTMAS TREE CLUSTER (NGC 2264) ⊙

Betelgeuse

Procyon

CANIS MINOR

2244 Rosette

ORION

HYDRA

δ

ξ

MONOCEROS

β   γ

α

M50

2506

PUPPIS

LEPUS

HOW TO FIND NCG 2264

First, locate the constellations Canis Minor and Orion. The cluster is midway between Procyon in Canis Minor and Betelgeuse in Orion. It is below the foot star, Al Zirr, in Gemini.

CANIS MAJOR

CHAPTER EIGHT

NEBULAE

Our galaxy is ever-changing, although the changes happen at a slow pace in terms of human perception. Over the course of millions of years, stars are born, go through a life cycle, and die.

In this process, a lot of gaseous matter is formed and reformed. And this matter can be concentrated into something called nebulae. Through the telescope, these nebulae appear as fluffy, cotton-ball-like structures in a variety of shapes and colors.

THE TYPES OF NEBULAE

Nebulae are categorized into four major types: diffuse, planetary, dark, and supernovae remnants. You will also look at a diffuse nebula that is currently birthing stars. In this type of nebula, matter is drawn together to incubate and bring to life new stars. The Orion Nebula is a good example of this.

Diffuse Nebulae. The most common type, a diffuse nebula presents a light and wispy cloud-like structure without a well-defined shape, almost as if its ever-lightening borders just extend out forever. You can see diffuse nebulae because they either emit their own light (emission nebulae) or they reflect light from nearby stars (reflection nebulae). Diffuse nebulae are also the birthplace of stars. In diffuse nebulae, the gases are being slowly pulled together by gravitational forces. Over time, this matter forms into a star. Some nebulae that we see now have newly formed stars and some of the nebulae we see now have not yet formed stars.

Planetary Nebulae. A planetary nebula, which often looks like a donut, is very different from a diffuse nebula. They are formed when a nova, or tremendous explosion of a star, occurs, which creates an even shell of matter that extends out in space from the remains of the star. Early astronomers couldn't resolve them well with their telescopes, so they called them planetary nebulae because they had width and were well defined, just like a planet.

Dark Nebulae. This type of nebula is very similar in shape and structure to diffuse nebulae, with the exception that they don't emit or reflect any light. This causes them to block the light from objects behind them, so they appear to be just a dark area, but are often very rich and rewarding to observe.

Supernova Remnant Nebulae. Many nebulae are the result of a star experiencing a nova. But, on rare occasions, a supernova will occur, which is a violent explosion within a star that has much more force than a nova. Rather than a uniform expansion of matter into space like with a planetary nebula, supernova remnant nebulae appear in erratic and dramatic shapes. The Crab Nebula is a good example of this.

40. THE ORION NEBULA (M42)

OBJECT TYPE: Diffuse emission nebula

CONSTELLATION: Orion

APPARENT MAGNITUDE: +4

COORDINATES: 05h 35m 17.3s, -05° 23' 28"

SEASON: December through March; best in January

DIFFICULTY: Easy

This is one of the best sights among all the nebulae. It is extremely easy to find because of its brightness and its location among the stars of Orion.

If you are using a small telescope, you will not see the Orion Nebula as nicely as I have shown in the circular telescope view here, but I just couldn't resist giving you a nice picture of it. It is simply too beautiful. Generally, it will reveal itself with a little bit of detail and no color. It will look like a cotton ball that has been pulled and teased a bit.

In this cloudy formation, large sections are being pulled together by gravity. These areas are in various stages of star and planetary formation. The nebulae in which this process occurs are referred to as stellar nurseries.

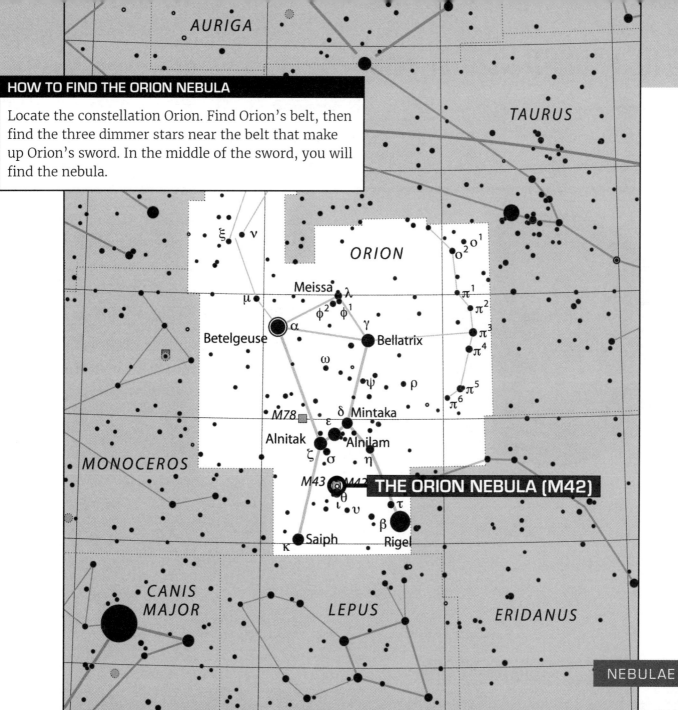

HOW TO FIND THE ORION NEBULA

Locate the constellation Orion. Find Orion's belt, then find the three dimmer stars near the belt that make up Orion's sword. In the middle of the sword, you will find the nebula.

AURIGA

TAURUS

ORION

Meissa λ

μ

φ² φ¹

γ

α

Betelgeuse

Bellatrix

ω

ψ

ρ

M78

δ Mintaka

ε

Alnitak

Alnilam

ζ σ η

M43 M42

THE ORION NEBULA (M42)

θ

ι υ

τ

β

κ Saiph

Rigel

π¹

π²

π³

π⁴

π⁵

π⁶

o² o¹

MONOCEROS

CANIS MAJOR

LEPUS

ERIDANUS

41. THE LAGOON NEBULA (M8)

OBJECT TYPE: Diffuse emission nebula

CONSTELLATION: Sagittarius

APPARENT MAGNITUDE: +6.0

COORDINATES: 18h 03m 37s, -24° 23' 12"

SEASON: June through August

DIFFICULTY: Medium

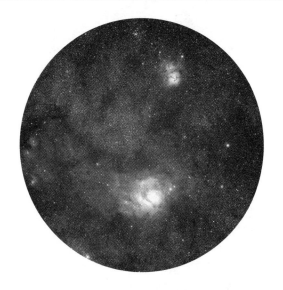

A possible second to the Orion Nebula, this one is trickier to find. However, on good nights, it is visible to the naked eye. The picture here shows it within the thick cloud formations of Sagittarius. It is the bright and diffuse section in the center.

Because this is an emission nebula, it is illuminated by its own light. It doesn't reflect light from nearby stars.

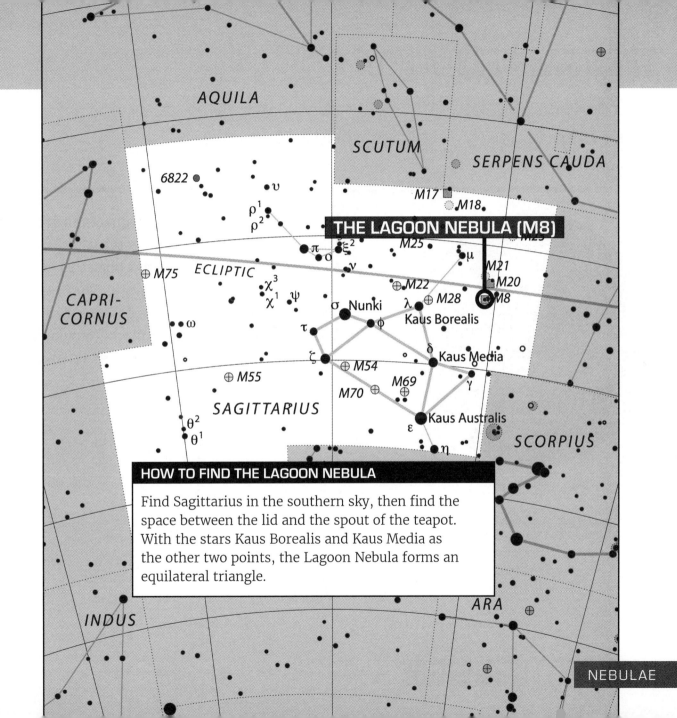

THE LAGOON NEBULA (M8)

HOW TO FIND THE LAGOON NEBULA

Find Sagittarius in the southern sky, then find the
space between the lid and the spout of the teapot.
With the stars Kaus Borealis and Kaus Media as
the other two points, the Lagoon Nebula forms an
equilateral triangle.

42. THE SWAN NEBULA [M17]

OBJECT TYPE: Diffuse emission nebula

CONSTELLATION: Sagittarius

APPARENT MAGNITUDE: +6.0

COORDINATES: 18h 20m 26s, -16° 10' 36"

SEASON: June through August

DIFFICULTY: Medium

This one is often a favorite of amateur astronomers, and through the telescope it actually looks like a swan. But, while it is a good object to observe, it can be a bit tricky to initially find. The Swan Nebula also goes by several other names, including the Omega Nebula, the Checkmark Nebula, and the Horseshoe Nebula. You decide which name best fits its appearance in your telescope.

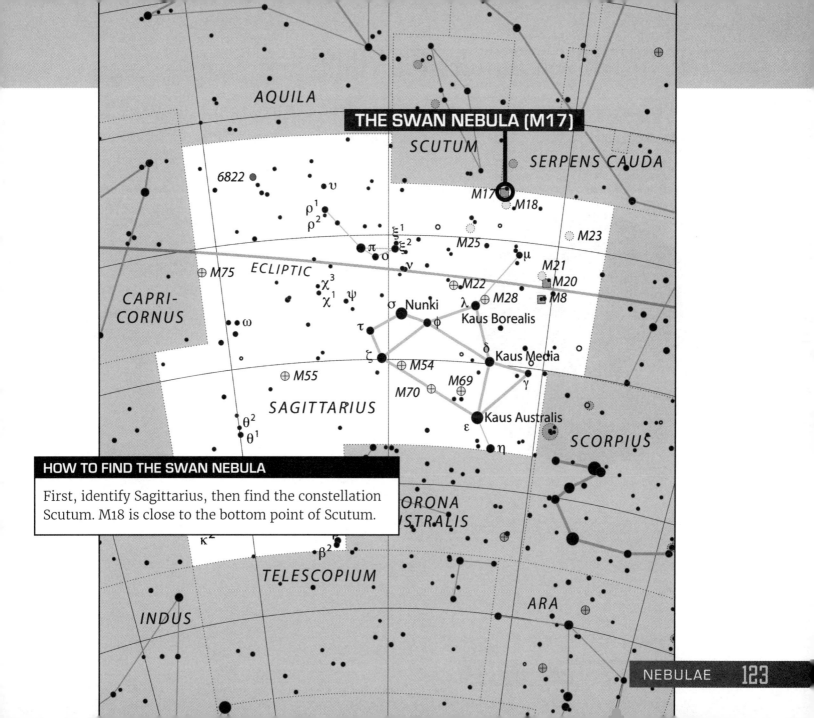

THE SWAN NEBULA (M17)

AQUILA

SCUTUM

SERPENS CAUDA

6822

υ

ρ¹
ρ²

M17 · M18

ξ¹

M23

π · ξ²

M25

M21

ECLIPTIC

⊕ M75

χ³

ν

μ

M20

o

CAPRI-
CORNUS

χ¹ ψ

σ Nunki

M22

λ

M28

M8

ω

φ

Kaus Borealis

τ

δ Kaus Media

ζ

⊕ M54

γ

⊕ M55

M70 M69

SAGITTARIUS

Kaus Australis

θ²

ε

θ¹

η

SCORPIUS

HOW TO FIND THE SWAN NEBULA

First, identify Sagittarius, then find the constellation
Scutum. M18 is close to the bottom point of Scutum.

CORONA
AUSTRALIS

κ²

β²

TELESCOPIUM

ARA

INDUS

43. THE SATURN NEBULA (NGC 7009)

OBJECT TYPE: Planetary nebula

CONSTELLATION: Aquarius

APPARENT MAGNITUDE: +8.0

COORDINATES: 18h 20m 26s, -16° 10' 36"

SEASON: June through August

DIFFICULTY: Difficult

This was called the Saturn Nebula by early astronomers because of its very strong resemblance to the planet, as you can see in the image here. With your telescope, it may appear football shaped. It can be a challenge to locate because it is dim. But, it is well worth the hunt, because once you bring up the magnification on most small telescopes, you will be able to see a slight greenish hue to this nebula.

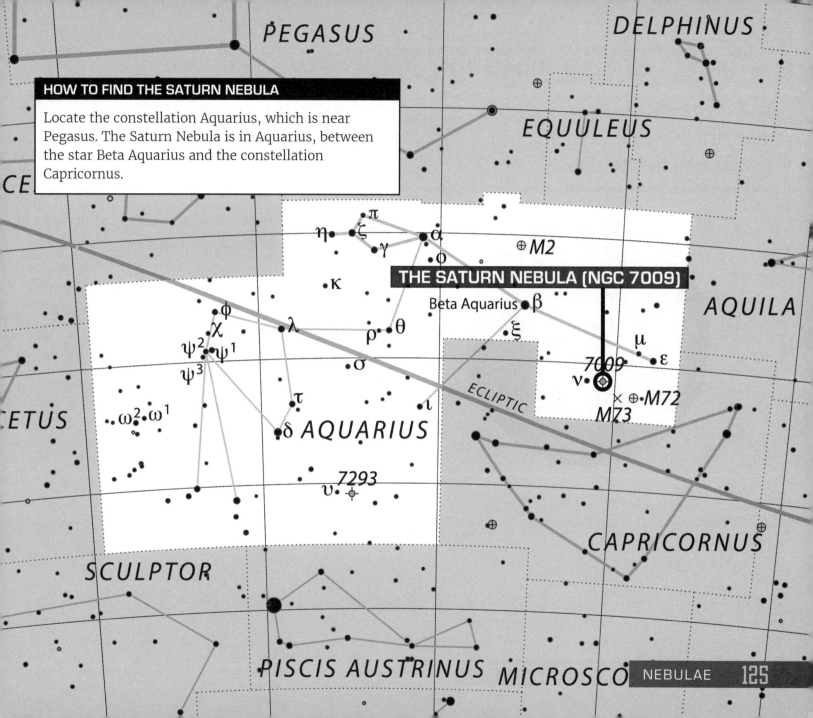

PEGASUS

DELPHINUS

HOW TO FIND THE SATURN NEBULA

Locate the constellation Aquarius, which is near
Pegasus. The Saturn Nebula is in Aquarius, between
the star Beta Aquarius and the constellation
Capricornus.

EQUULEUS

CE

η ζ ζ π
γ α
φ
⊕ M2

κ

THE SATURN NEBULA (NGC 7009)

Beta Aquarius • β

AQUILA

CE

φ
χ λ
ψ² ψ¹
ψ³
ρ θ
ξ

μ
ε

7009
ν ⊙

ECLIPTIC
× ⊕ M72
M73

ω² ω¹
σ
τ
δ AQUARIUS
ι

7293
υ •⊕

CETUS

CAPRICORNUS

SCULPTOR

PISCIS AUSTRINUS MICROSCO

44. JUPITER'S GHOST (NGC 3242)

OBJECT TYPE: Planetary nebula

CONSTELLATION: Hydra

APPARENT MAGNITUDE: +8.6

COORDINATES: 10h 24m 46.1s, -18° 38' 32.6"

SEASON: March through May; best in April

DIFFICULTY: Difficult

This nebula is referred to as Jupiter's Ghost because of its resemblance in size and shape to the planet. It is also called the Eye Nebula because with larger telescopes, it looks like a bright star in the center with a halo-like shell around it, the whole of which resembles an eye. On good viewing nights, you may be able to see its bluish-green tint even with a small telescope.

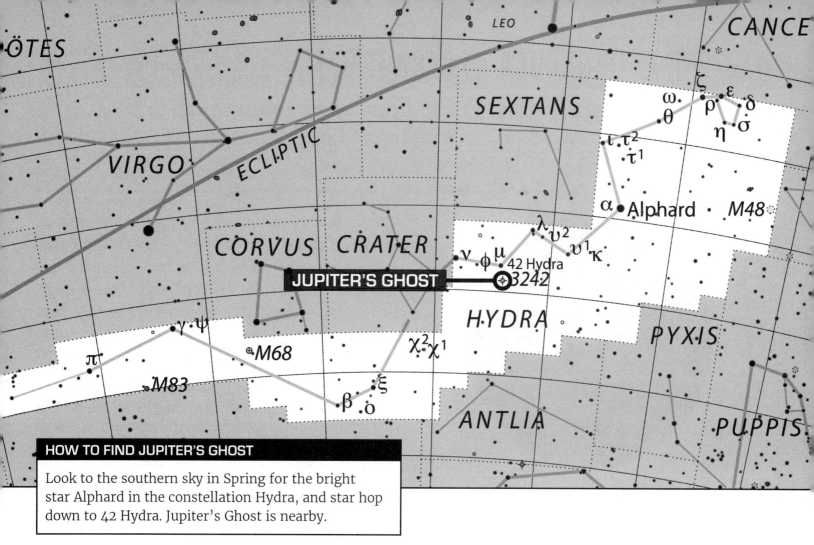

ÖTES

LEO

CANCE

VIRGO

ECLIPTIC

SEXTANS

ω θ ζ ρ ε δ
η σ
ι τ² τ¹

α Alphard · M48

CORVUS CRATER

ν φ μ υ² λ
42 Hydra
υ¹ κ

JUPITER'S GHOST ———●⊙3242

HYDRA

PYXIS

γ ψ

χ² χ¹

π

M68

M83

ξ

β ο

ANTLIA

PUPPIS

HOW TO FIND JUPITER'S GHOST

Look to the southern sky in Spring for the bright
star Alphard in the constellation Hydra, and star hop
down to 42 Hydra. Jupiter's Ghost is nearby.

45. THE RING NEBULA IN LYRA (M57)

OBJECT TYPE: Planetary nebula

CONSTELLATION: Lyra

APPARENT MAGNITUDE: +8.8

COORDINATES: 18h 53m 35.08s, +33° 01' 45.03"

SEASON: May through October; best in summer

DIFFICULTY: Easy

This planetary nebula is not overly impressive, but it's a really good choice if you want to easily find a planetary nebula. It is located in an easy-to-find spot not far from the very bright star of Vega. You will have very little trouble finding this deep space object. It's also an interesting test for your telescope. With smaller telescopes, it will look like a round shape. With a little bit of a bigger telescope or with really good viewing, you should be able to discern the hole in the middle, thus making it a ring.

This was the first planetary nebula I found in my telescope when I was a teenager. Many years later, I still remember how it looked and how I felt having found it. Every summer, I enjoy returning to it just to get that sense of nostalgia.

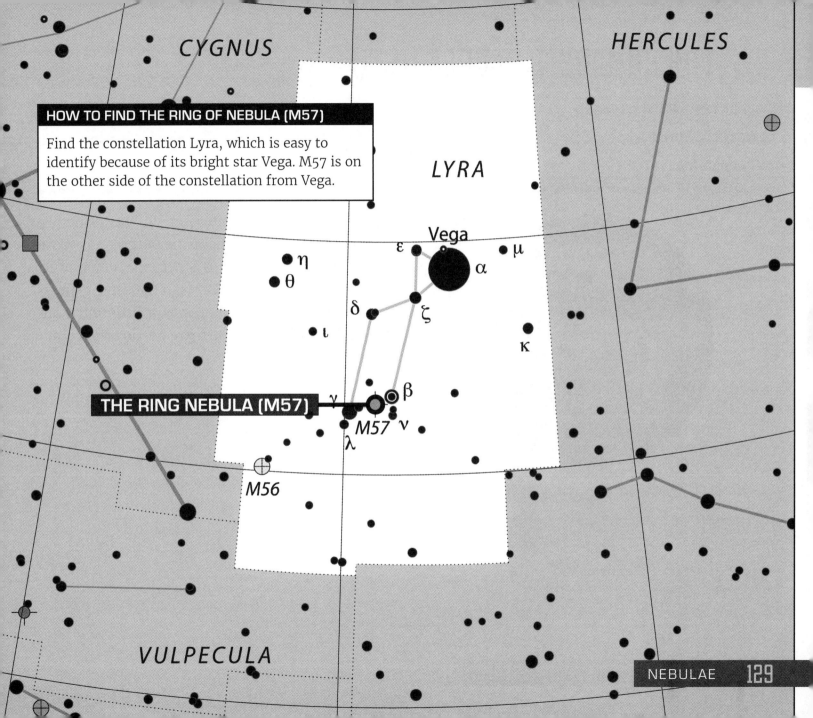

CYGNUS

HERCULES

HOW TO FIND THE RING OF NEBULA (M57)

Find the constellation Lyra, which is easy to
identify because of its bright star Vega. M57 is on
the other side of the constellation from Vega.

LYRA

Vega

ε

η

θ

μ

α

δ

ζ

ι

κ

THE RING NEBULA (M57)

γ

β

M57

ν

λ

M56

VULPECULA

46. THE DUMBBELL NEBULA (M27)

OBJECT TYPE: Planetary nebula

CONSTELLATION: Vulpecula

APPARENT MAGNITUDE: +7.5

COORDINATES: 19h 59m 36.34s, +22° 43' 16.09"

SEASON: July through November; best in September

DIFFICULTY: Medium

This is one of the brightest of all the planetary nebulae. It also goes by the name of the Apple Core nebula. If you have trouble seeing the shape of it, try using averted vision. M27 is in the same general area of the sky as M57 but it is a little bit more challenging to find because there are no bright stars or constellations close by.

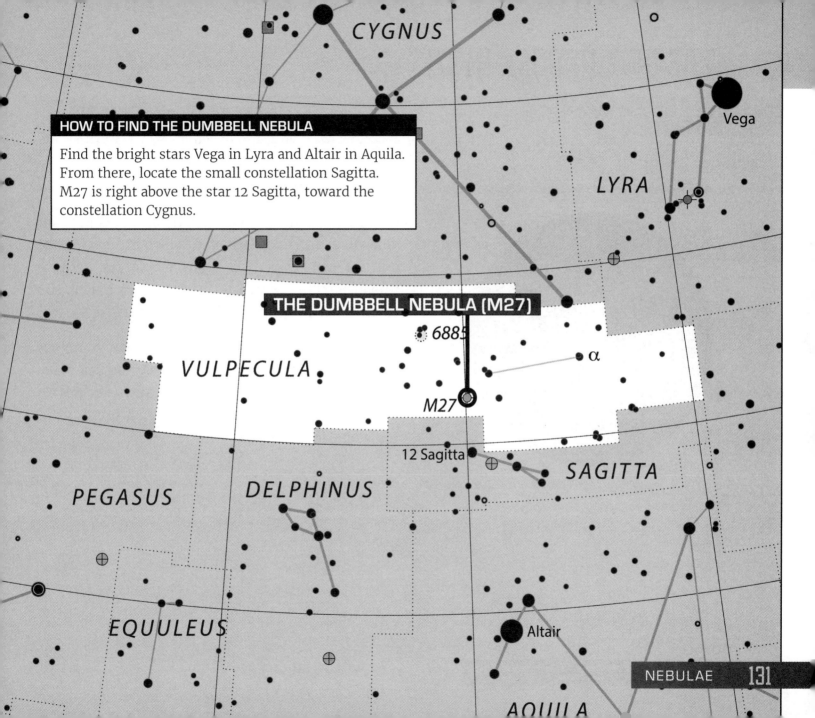

CYGNUS

Vega

LYRA

HOW TO FIND THE DUMBBELL NEBULA

Find the bright stars Vega in Lyra and Altair in Aquila. From there, locate the small constellation Sagitta. M27 is right above the star 12 Sagitta, toward the constellation Cygnus.

THE DUMBBELL NEBULA (M27)

6885

VULPECULA

α

M27

12 Sagitta

SAGITTA

PEGASUS

DELPHINUS

EQUULEUS

Altair

AQUILA

47. THE OWL NEBULA (M97)

OBJECT TYPE: Planetary nebula
Constellation: Ursa Major

APPARENT MAGNITUDE: +9.9

COORDINATES: 11h 14m 47.73s, +55° 01' 08.50"

SEASON: Available year-round for most of North America

DIFFICULTY: Medium

The owl nebula is visible with most telescopes, but you will need a telescope of at least 8 inches to make out the owl's eyes. It is a dim nebula, but because it is located in the constellation of Ursa Major, you should be able to find it without too much difficulty.

The telescope view you see here has a higher magnification and a larger telescope. You can make out the owl's eyes. If you can see the owl's eyes with your telescope, then you must have a pretty good telescope and good dark skies!

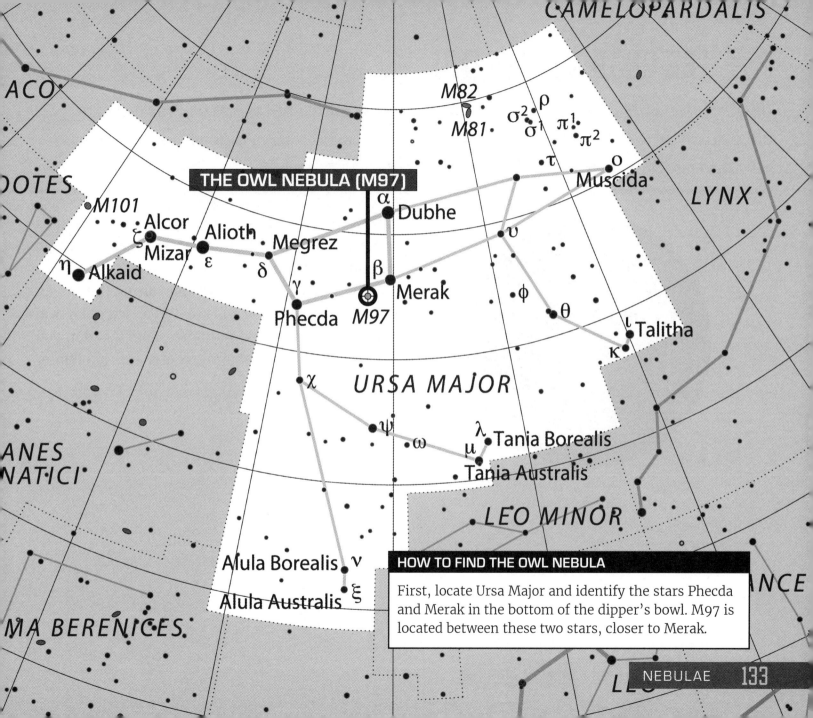

CAMELOPARDALIS

ACO

DOTES

M101

THE OWL NEBULA (M97)

α Dubhe

Alcor
ζ
Alioth
Mizar
ε
δ
Megrez
η
Alkaid
γ
β
Phecda
M97
Merak

M82
M81
σ² ρ
σ¹ π¹
π²
τ
Muscida

LYNX

υ

φ
θ
ι Talitha
κ

χ
URSA MAJOR

ψ
ω
λ
μ Tania Borealis
Tania Australis

LEO MINOR

ANES
NATICI

Alula Borealis ν
Alula Australis ξ

MA BERENICES

HOW TO FIND THE OWL NEBULA

First, locate Ursa Major and identify the stars Phecda and Merak in the bottom of the dipper's bowl. M97 is located between these two stars, closer to Merak.

ANCE

LEO

48. THE CRAB NEBULA (M1)

OBJECT TYPE: Supernova remnant

CONSTELLATION: Taurus

APPARENT MAGNITUDE: +8.4

COORDINATES: 05h 34m 31.94s, +22° 00' 52.2"

SEASON: November through April; best in January

DIFFICULTY: Medium

This is an interesting object because we know it as a supernova remnant. It was observed and recorded by Chinese astronomers in the year 1054, and it was so bright that it was even visible during the day. In a small telescope, you will see it as an irregular oval shape. If you move up to higher magnification you should be able to see that the center area is a lighter shade.

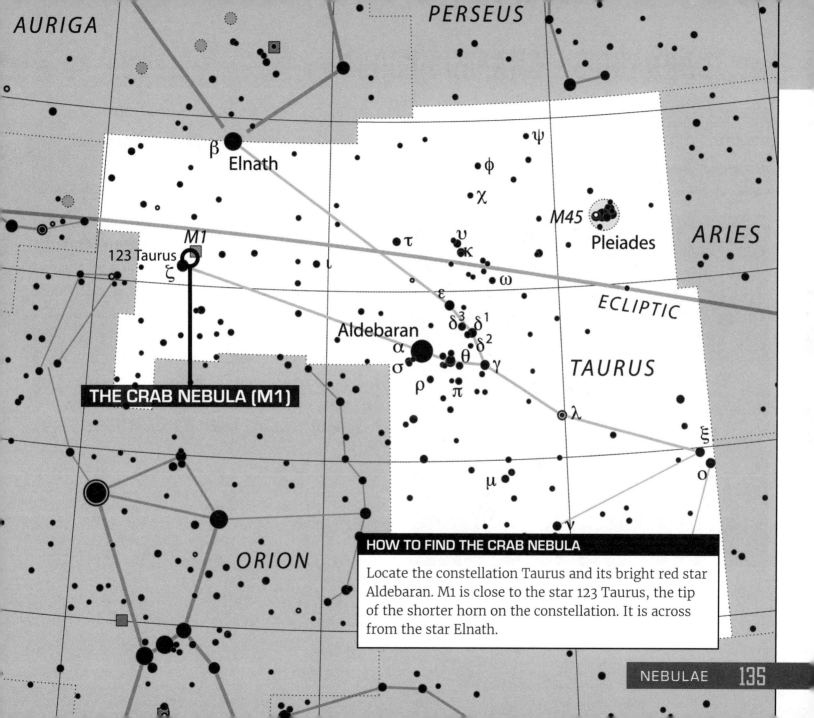

THE CRAB NEBULA (M1)

HOW TO FIND THE CRAB NEBULA

Locate the constellation Taurus and its bright red star Aldebaran. M1 is close to the star 123 Taurus, the tip of the shorter horn on the constellation. It is across from the star Elnath.

49. BARNARD'S E-SHAPED NEBULA

OBJECT TYPE: Dark nebula

CONSTELLATION: Aquila

APPARENT MAGNITUDE: N/A

COORDINATES: 19h 40m 42s, +10° 57' 00"

SEASON: July through October; best in August

DIFFICULTY: Medium

Edward Emerson Barnard was an astronomer who compiled a list of dark nebulae in 1919. This list is called the Barnard Catalog. If you are curious about dark nebulae, this catalog is a great source for learning more. The original 1919 catalog has 182 items on it. He continued to add to the list until it contained 369 objects. This E-shaped nebula is listed as items 142 and 143 in the catalog. 142 is the lower part, which is shaped like an underscore, and 143 is the upper part, which is shaped like a C.

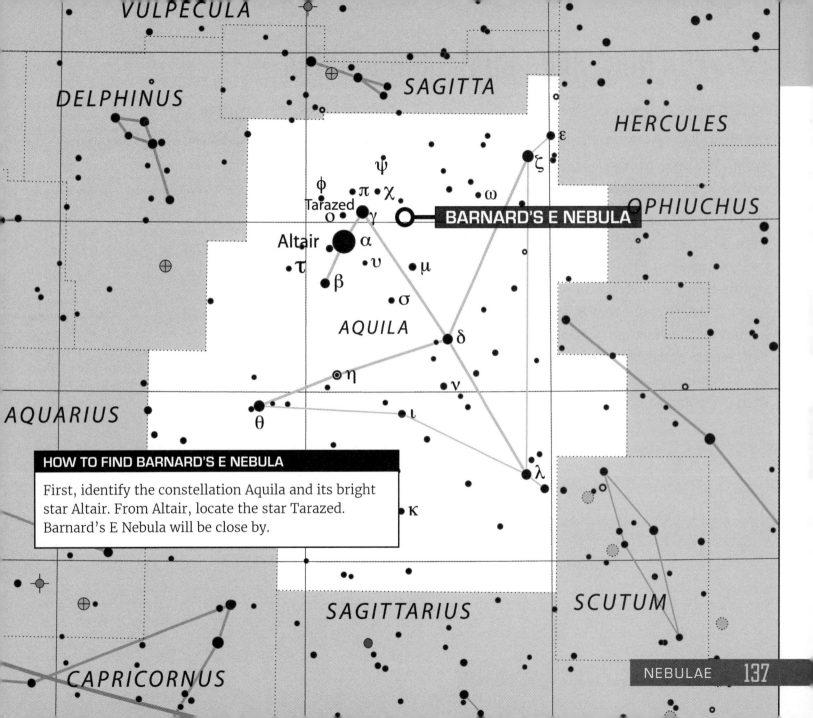

VULPECULA

DELPHINUS

SAGITTA

HERCULES

ε

ζ

ψ

φ

π χ

ω

OPHIUCHUS

Tarazed

ο γ

BARNARD'S E NEBULA

Altair

α

τ

υ

μ

β

σ

AQUILA

δ

η

ν

AQUARIUS

θ

ι

λ

HOW TO FIND BARNARD'S E NEBULA

First, identify the constellation Aquila and its bright
star Altair. From Altair, locate the star Tarazed.
Barnard's E Nebula will be close by.

κ

SCUTUM

SAGITTARIUS

CAPRICORNUS

50. THE NORTHERN COALSACK NEBULA

OBJECT TYPE: Dark nebula

CONSTELLATION: Cygnus

APPARENT MAGNITUDE: N/A

COORDINATES: 21h 08m 00s, +47° 36' 00"

SEASON: July through November; best in September

DIFFICULTY: Easy

This dark nebula is the beginning portion of the Dark Rift, which is a part of the Milky Way you had looked at in Chapter 6. You may think of it as simply being an absence of anything, but it is not. It is a gaseous cloud that eclipses the light of stars behind it. The Northern Coalsack is easily found on dark nights and is best viewed either with the naked eye or binoculars. Once you have located it, you can then use your telescope to hunt around its insides.

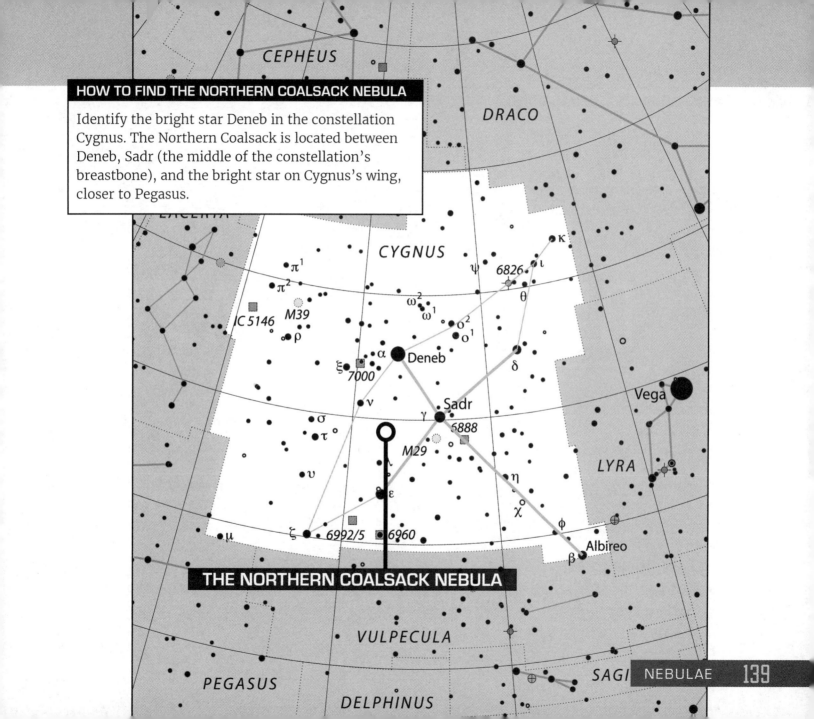

HOW TO FIND THE NORTHERN COALSACK NEBULA

Identify the bright star Deneb in the constellation Cygnus. The Northern Coalsack is located between Deneb, Sadr (the middle of the constellation's breastbone), and the bright star on Cygnus's wing, closer to Pegasus.

THE NORTHERN COALSACK NEBULA

CHAPTER NINE

OTHER GALAXIES

Once you step outside our Milky Way, you enter the very deep and very dark areas of space. The distance between objects is so large that it is difficult to fully comprehend, and using a small telescope, the only things you can see at these distances are other galaxies.

This is because they are composed of millions of stars, and somehow their combined light manages to cross extreme distances to reach us. Don't let that explanation of distance discourage you: There are some fantastic galaxies you can look at, and many of them are extremely easy to find.

This will be a very fun chapter because you are going to look at galaxies in a wide variety of positions including from the top, from the side, and from a three-quarter angle. And many of them have terrific and memorable names that describe to us how they look through a telescope.

TIPS FOR OBSERVING GALAXIES WITH YOUR TELESCOPE

Hundreds of galaxies are visible with a 4-inch telescope, and of these hundreds, several dozen are well worth viewing because they will display some interesting shapes and formations other than just dots of light.

The biggest thing to consider when viewing galaxies is the darkness of the sky. Some galaxies have something called low surface brightness, meaning the light of the galaxy appears dimmer than the surrounding area. A very dark sky is needed to give you contrast when viewing them.

Use and practice the technique of averted vision when observing galaxies. You will notice a big difference in what you can see. Additionally, it is usually very rewarding to return back to a galaxy later in the evening. This is for two reasons. First, the night sky is probably darker because the sun has moved further around to the other side of the Earth. Second, you will have been out in the dark for a much longer period of time and your eyes will be more sensitive to light.

51. THE ANDROMEDA GALAXY (M31)

OBJECT TYPE: Spiral galaxy

CONSTELLATION: Andromeda

APPARENT MAGNITUDE: +3.44

COORDINATES: 00h 42m 44.3s, +41° 16' 9"

SEASON: August through November; best in October

DIFFICULTY: Easy

Also known as the Great Spiral Galaxy, Andromeda is composed of hundreds of millions of stars. Some estimates put it at a trillion stars.

This spiral galaxy is our nearest neighbor in space beyond the Milky Way galaxy, and it gives us a dramatic view of what a spiral galaxy is all about. It is easy to find and very bright in terms of deep space objects. On a good night, you can see it with the naked eye, which makes it very easy to locate. It is also the most distant object you can see with the naked eye. Find it, and you are looking at an object that is over 2 million light years away.

The three-quarter view you'll get of this galaxy is simply astonishing. By observing it, you'll get an excellent idea of how a spiral galaxy is shaped.

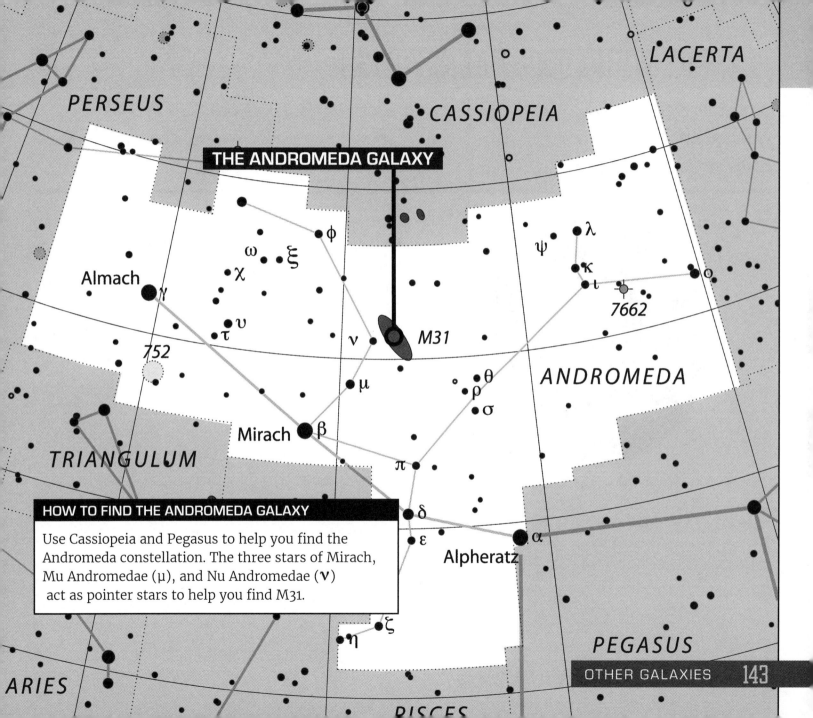

PERSEUS

LACERTA

CASSIOPEIA

THE ANDROMEDA GALAXY

φ

ω ξ

χ

Almach

γ

υ

τ

752

ν

M31

μ

Mirach β

π

ψ

λ

κ

ι

7662

θ

ρ

σ

ANDROMEDA

Θ

TRIANGULUM

δ

ε Alpheratz α

HOW TO FIND THE ANDROMEDA GALAXY

Use Cassiopeia and Pegasus to help you find the
Andromeda constellation. The three stars of Mirach,
Mu Andromedae (μ), and Nu Andromedae (ν)
 act as pointer stars to help you find M31.

ζ

η

PEGASUS

ARIES

PISCES

52. THE WHIRLPOOL GALAXY [M51]

OBJECT TYPE: Spiral galaxy

CONSTELLATION: Canes Venatici

APPARENT MAGNITUDE: +8.4

COORDINATES: 13h 29m 52.7s, +47° 11' 43"

SEASON: Available year-round for most of North America

DIFFICULTY: Medium

This is a beautiful spiral galaxy with a second, smaller galaxy spinning off of it. But, they can be tricky to see without dark skies because of their low surface brightness. One interesting thing about this galaxy pair is that you get a full top view of it rather than a view from the side or from an angle. The two of them together are simply referred to as M51, but the larger one is M51a and the smaller one is M51b.

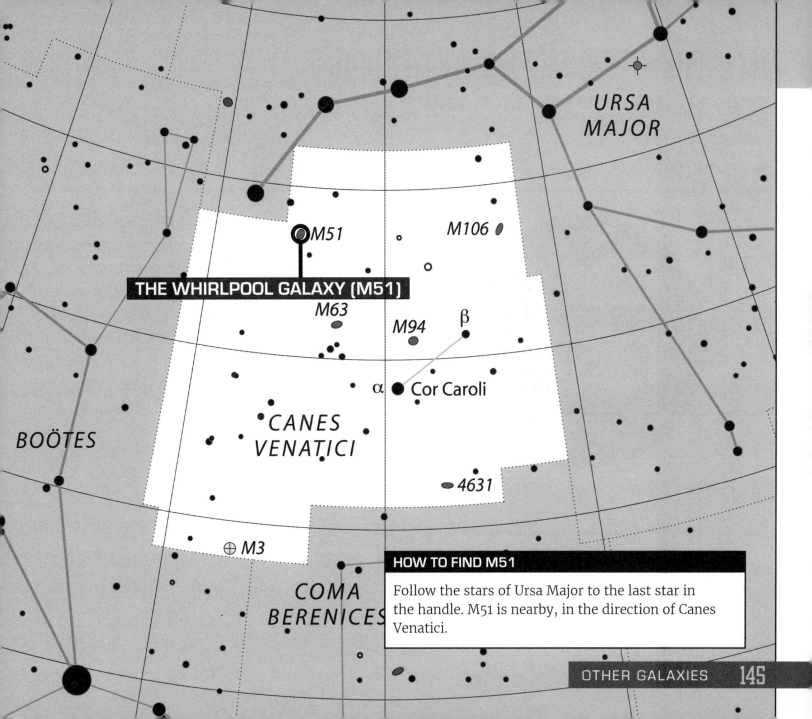

URSA MAJOR

M51

M106

THE WHIRLPOOL GALAXY (M51)

M63

M94

β

α Cor Caroli

CANES VENATICI

BOÖTES

4631

⊕ M3

COMA BERENICES

HOW TO FIND M51

Follow the stars of Ursa Major to the last star in the handle. M51 is nearby, in the direction of Canes Venatici.

53. THE SOMBRERO GALAXY (M104)

OBJECT TYPE: Spiral galaxy

CONSTELLATION: Virgo

APPARENT MAGNITUDE: +8.98

COORDINATES: 12h 39m 59.4s, -11° 37' 23"

SEASON: April through July; best in May

DIFFICULTY: Medium

The Sombrero Galaxy is a spiral galaxy that gives us an excellent side view. And it has a prominent black line cutting right through the spiral edge of it. The black line is a thick dust lane that gives it the appearance of a sombrero. You can usually make that feature out with a dark sky and 6-inch telescope or larger.

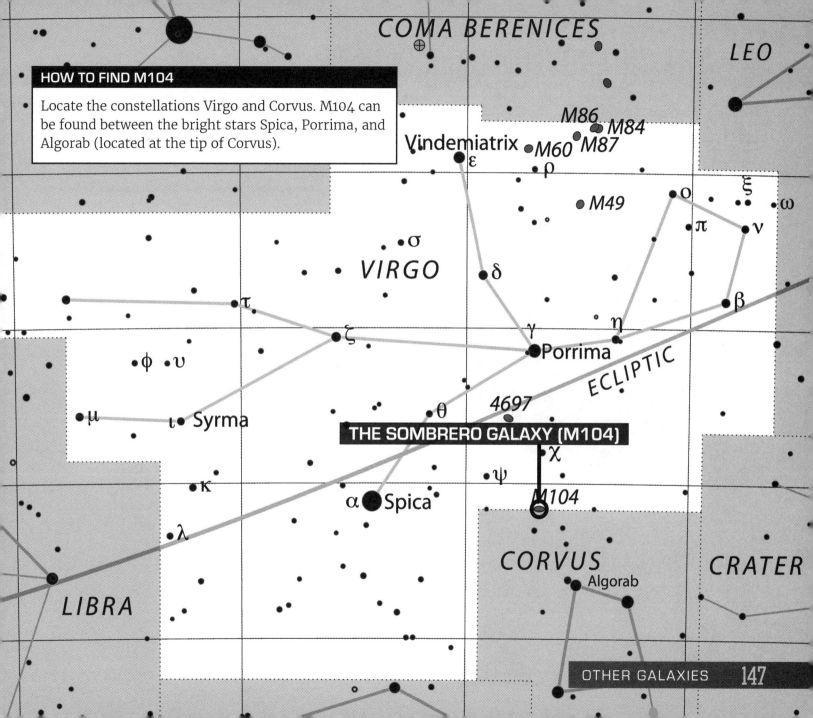

COMA BERENICES

LEO

HOW TO FIND M104

Locate the constellations Virgo and Corvus. M104 can be found between the bright stars Spica, Porrima, and Algorab (located at the tip of Corvus).

M86 M84
Vindemiatrix M60 M87
ε ρ

M49
o ξ
ω
π ν

σ
VIRGO δ

τ ζ γ η β

Porrima
ECLIPTIC

φ υ

μ ι Syrma θ 4697

THE SOMBRERO GALAXY (M104)

χ
ψ

κ M104

α Spica
λ

CORVUS CRATER
Algorab

LIBRA

54. THE TRIANGULUM GALAXY [M33]

OBJECT TYPE: Spiral galaxy

CONSTELLATION: Triangulum

APPARENT MAGNITUDE: +5.72

COORDINATES: 01h 33m 50.02s, +30° 39' 36.7"

SEASON: November through March; best in December

DIFFICULTY: Difficult

This is the second closest galaxy to our Milky Way, with the closest being the Andromeda Galaxy. It is a relatively bright deep space object, but its surface brightness is low, so it is best found with binoculars or a low-power eyepiece. Light pollution can have a strong effect on whether you can see it clearly.

HOW TO FIND M33

The Triangulum Galaxy is between the constellation Triangulum and the bright star Mirach in the constellation Andromeda.

ANDROMEDA

PERSEUS

Mirach

δ

β

γ

ε

M33

TRIANGULUM

α

THE TRIANGULUM GALAXY (M33)

Hamal

PISCES

ARIES

55. THE SOUTHERN PINWHEEL GALAXY (M83)

OBJECT TYPE: Barred spiral galaxy

CONSTELLATION: Hydra

APPARENT MAGNITUDE: +7.54

COORDINATES: 13h 37m 00.9s, -29° 51' 57"

SEASON: March through May; best in April

DIFFICULTY: Medium

This is a wonderful galaxy to observe with your telescope, but it is in a rather empty region of the night sky without a lot of guide stars or constellations to help you. But it's well worth viewing. This galaxy is a barred spiral galaxy, which is a type of spiral galaxy that has a bar-like bulge that extends from the central core to the spiral arms, unlike typical spiral galaxies that curve from the center to the arms.

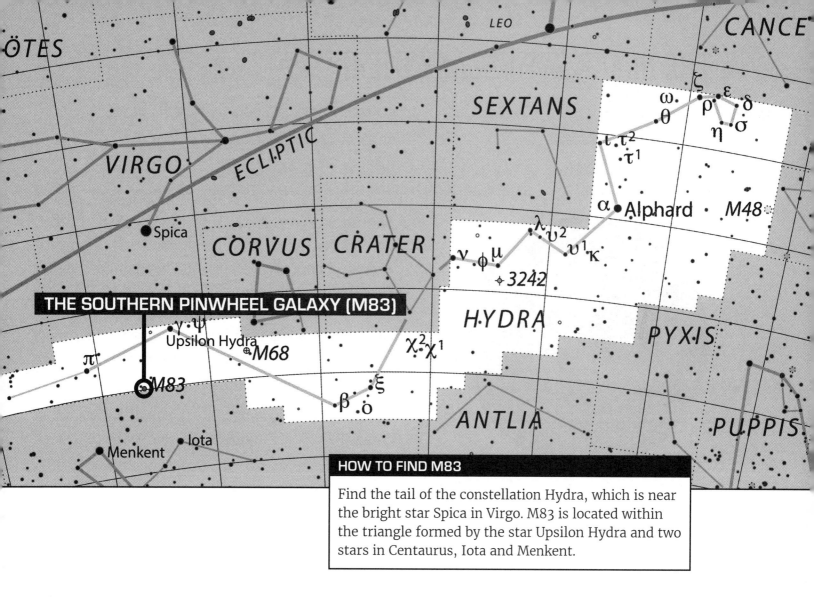

ÖTES

LEO

CANCE

SEXTANS

ω

ζ ρ ε δ

θ

η σ

ι τ²

τ¹

VIRGO

ECLIPTIC

α • Alphard • M48

λ υ²

• Spica

ν φ μ

υ¹ κ

CORVUS CRATER

⋄ 3242

HYDRA

PYXIS

THE SOUTHERN PINWHEEL GALAXY (M83)

γ ψ

Upsilon Hydra

χ² χ¹

π

⊕ M68

ξ

◎ M83

β

ο

ANTLIA

PUPPIS

Iota

Menkent

HOW TO FIND M83

Find the tail of the constellation Hydra, which is near
the bright star Spica in Virgo. M83 is located within
the triangle formed by the star Upsilon Hydra and two
stars in Centaurus, Iota and Menkent.

56. THE BLACK EYE GALAXY (M64)

OBJECT TYPE: Spiral galaxy

CONSTELLATION: Coma Berenices

APPARENT MAGNITUDE: +9.36

COORDINATES: 12h 56m 43.7s, +21° 40' 58"

SEASON: May through August; best in July

DIFFICULTY: Medium

The Black Eye Galaxy, sometimes called the Evil Eye Galaxy, has a dark band visible. This unusual look has happened because of a band of dark stellar dust that eclipses the starlight behind it. The dark band of dust rotates in the opposite direction of the galaxy, which has led to the theory that this galaxy is the result of a second and darker galaxy being absorbed by the brighter one.

You can find this galaxy in the small constellation Coma Berenices, named for the Hair of Queen Berenice.

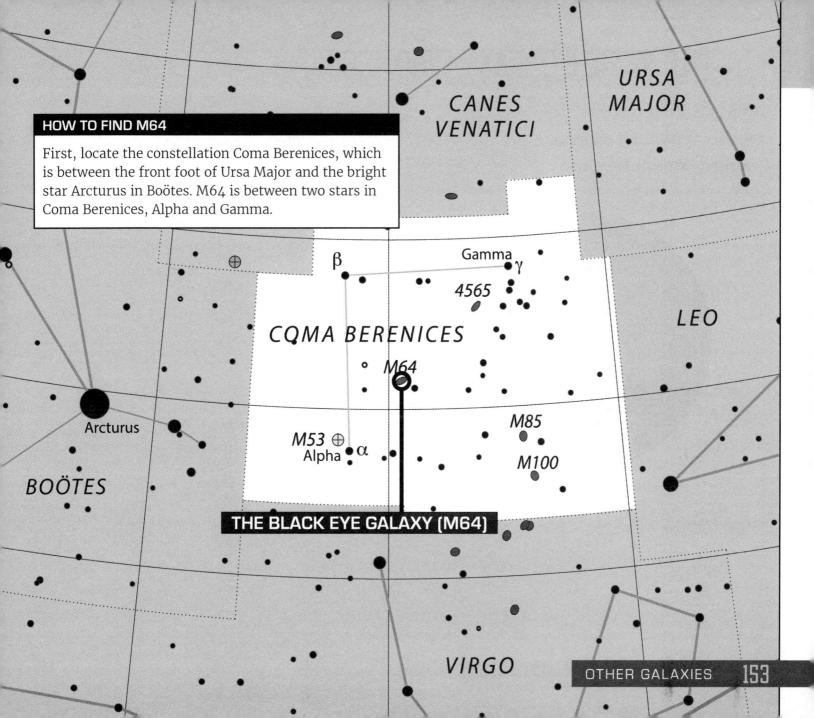

CANES VENATICI

URSA MAJOR

HOW TO FIND M64

First, locate the constellation Coma Berenices, which is between the front foot of Ursa Major and the bright star Arcturus in Boötes. M64 is between two stars in Coma Berenices, Alpha and Gamma.

β Gamma γ

4565

CQMA BERENICES

LEO

M64

M53 ⊕ α
Alpha

M85

M100

THE BLACK EYE GALAXY (M64)

Arcturus

BOÖTES

VIRGO

57. THE NEEDLE GALAXY (NGC 4565)

OBJECT TYPE: Edge-on spiral galaxy

CONSTELLATION: Coma Berenices

APPARENT MAGNITUDE: +10.4

COORDINATES: 12h 36m 20.8s, +25° 59' 16"

SEASON: May through August; best in July

DIFFICULTY: Medium

The side view of the Needle Galaxy, a spiral galaxy, is just remarkable, and because of this view it is aptly named.

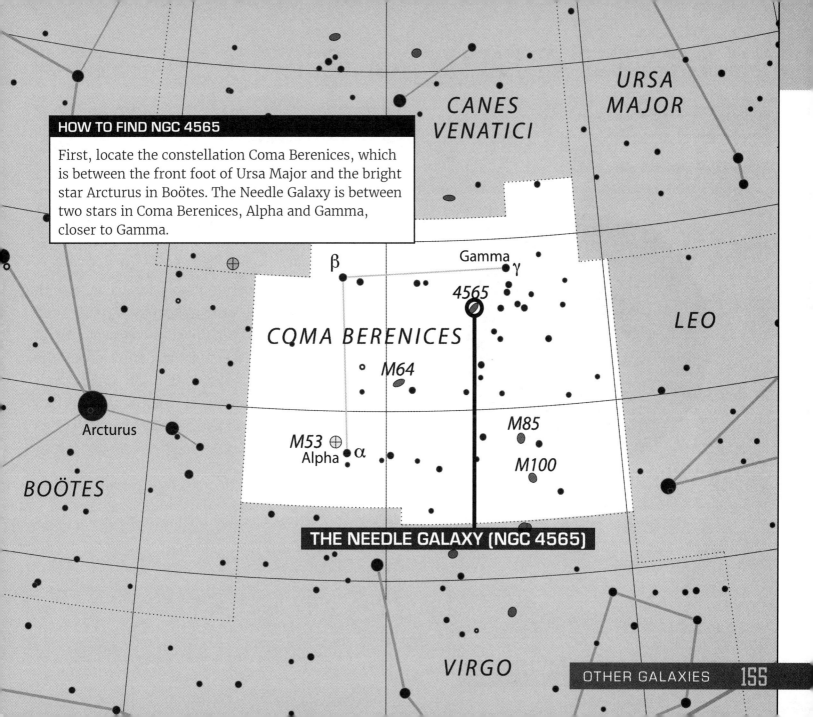

CANES VENATICI

URSA MAJOR

HOW TO FIND NGC 4565

First, locate the constellation Coma Berenices, which is between the front foot of Ursa Major and the bright star Arcturus in Boötes. The Needle Galaxy is between two stars in Coma Berenices, Alpha and Gamma, closer to Gamma.

β

Gamma
γ

4565

CQMA BERENICES

LEO

M64

M53

M85

Alpha
α

M100

THE NEEDLE GALAXY (NGC 4565)

Arcturus

BOÖTES

VIRGO

58. THE SCULPTOR GALAXY (NGC 253)

OBJECT TYPE: Spiral galaxy

CONSTELLATION: Sculptor

APPARENT MAGNITUDE: +8.0

COORDINATES: 00h 47m 33s, -25° 17' 18"

SEASON: November

DIFFICULTY: Medium

Also known as the Silver Dollar Galaxy because of its brightness, this magnitude 8 is one of the brighter galaxies within easy reach of small telescopes. This galaxy is also very suitable for viewing with binoculars. But it will only appear as a fuzzy spot. You will need a medium to large telescope of at least 10 inches to discern some definition of the spiral pattern. Even though small telescopes can't resolve the spiral pattern of this galaxy, it is a favorite among amateur astronomers because of its brightness.

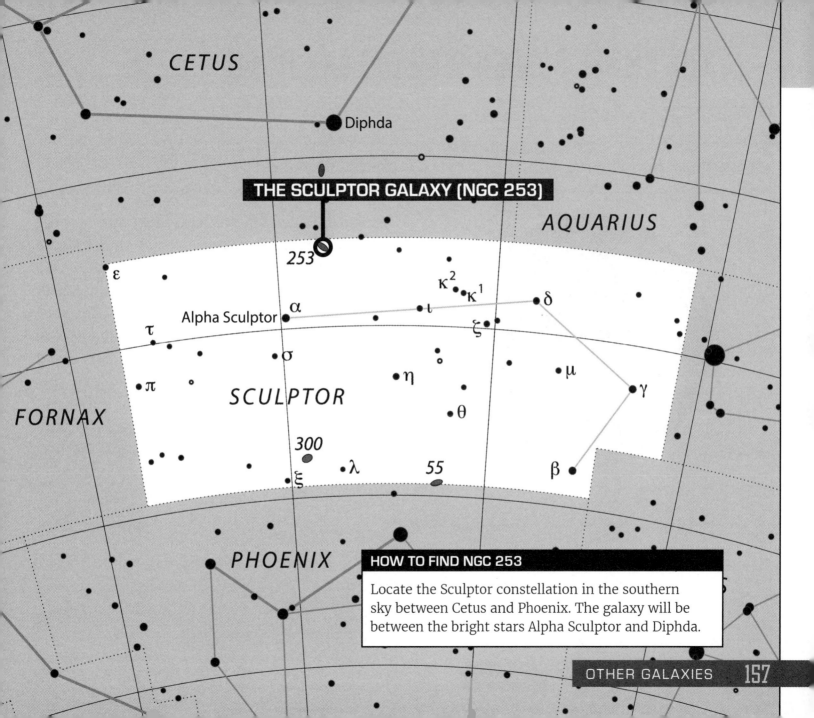

CETUS

Diphda

THE SCULPTOR GALAXY (NGC 253)

AQUARIUS

253

ε

α
Alpha Sculptor

τ

σ

π

FORNAX

SCULPTOR

κ² κ¹

ι

ζ

δ

η

μ

γ

θ

300

ξ

λ

55

β

PHOENIX

HOW TO FIND NGC 253

Locate the Sculptor constellation in the southern
sky between Cetus and Phoenix. The galaxy will be
between the bright stars Alpha Sculptor and Diphda.

59. THE SMALL GOLDEN RING GALAXY (M94)

OBJECT TYPE: Spiral galaxy

CONSTELLATION: Canes Venatici

APPARENT MAGNITUDE: +8.99

COORDINATES: 12h 50m 53.1s, +41° 07' 14"

SEASON: Available year-round for most of North America

DIFFICULTY: Medium

The Small Golden Ring Galaxy is also known as the Cat's Eye Galaxy. It gets its ring appearance because it is a galaxy with two sets of ring structures. One set of rings is near the core and blends with it. The second set of rings is farther out, with space between it and the core. It isn't exactly known why this galaxy has two sets of rings. There are several theories, which include the possibility that this galaxy is the result of one galaxy absorbing a second galaxy. It's also possible that it just appears this way because of the top-down view we see from Earth.

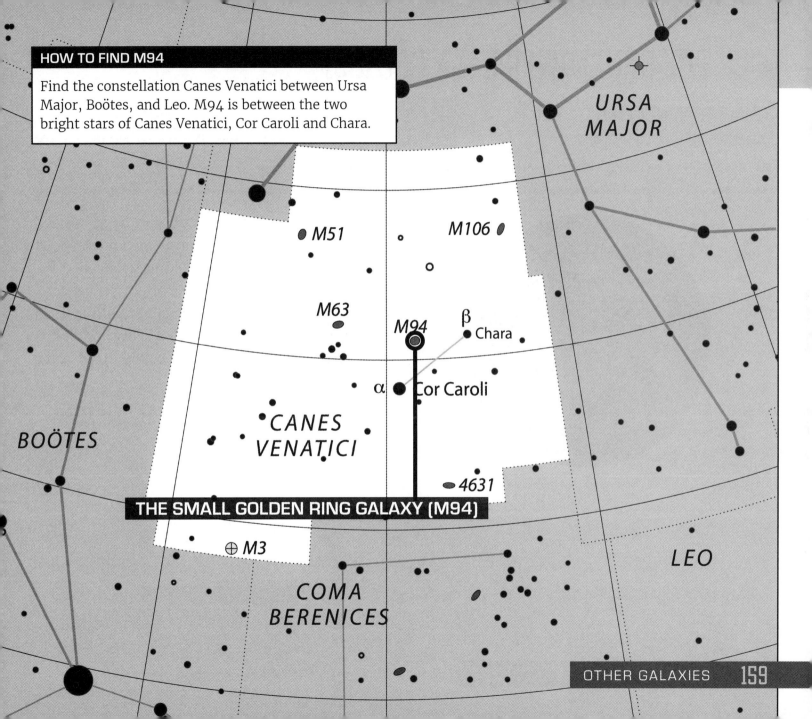

HOW TO FIND M94

Find the constellation Canes Venatici between Ursa Major, Boötes, and Leo. M94 is between the two bright stars of Canes Venatici, Cor Caroli and Chara.

URSA MAJOR

M51

M106

M63

M94   β
        Chara

α   Cor Caroli

CANES VENATICI

4631

BOÖTES

THE SMALL GOLDEN RING GALAXY (M94)

M3

COMA BERENICES

LEO

60. THE SUNFLOWER GALAXY (M63)

OBJECT TYPE: Spiral galaxy

CONSTELLATION: Canes Venatici

APPARENT MAGNITUDE: +9.3

COORDINATES: 13h 15m 49.3s, +42° 01' 45"

SEASON: Available year-round for most of North America

DIFFICULTY: Difficult

This galaxy is easy to find with binoculars or a small telescope, but you will need an 8-inch telescope or larger to make out the spiral arms. It's called the Sunflower Galaxy because of its bright yellow core, but the color is difficult to make out with a small telescope. On good nights, with a telescope that is 6 inches or larger, you should be able to get some good views.

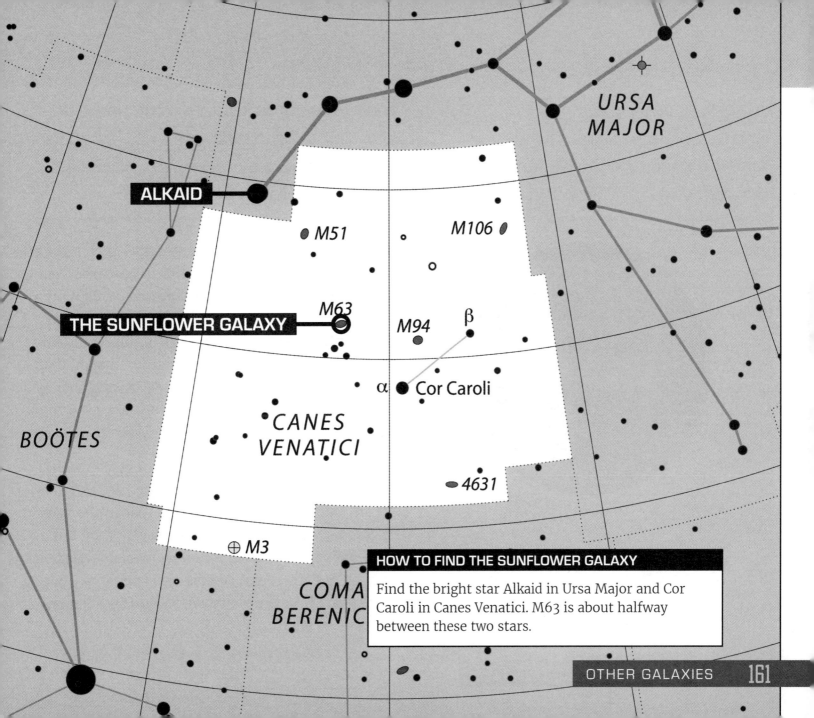

URSA
MAJOR

ALKAID

M51

M106

M63

THE SUNFLOWER GALAXY

M94

β

α Cor Caroli

BOÖTES

CANES
VENATICI

4631

M3

COMA
BERENIC

HOW TO FIND THE SUNFLOWER GALAXY

Find the bright star Alkaid in Ursa Major and Cor
Caroli in Canes Venatici. M63 is about halfway
between these two stars.

61. M81 AND M82

OBJECT TYPE: Double galaxy

CONSTELLATION: Ursa Major

APPARENT MAGNITUDE OF M81: +6.94

APPARENT MAGNITUDE OF M82: +8.41

COORDINATES OF M81: 09h 55m 33.2s, +69° 3' 55"

COORDINATES OF M82: 09h 55m 52.2s, +69° 40' 47"

SEASON: Available year-round for most of North America

DIFFICULTY: Medium

A double galaxy is a pair of galaxies that have a gravitational effect on each other. This makes them a single unit, similar to a double star, except on a much larger scale. These two galaxies are part of a group of galaxies called the M81 group, including several others that are significantly smaller than M81 and M82. The group is a nearby neighbor of ours, situated about 12 million light years away. And the two galaxies are 150,000 light years from each other.

Do not skip this pair! These galaxies are not very bright, but seeing them both together in the same telescopic view is a treat. And because they are in Ursa Major, they are in a readily available portion of the sky year-round. M81 is also known as Bodes Galaxy and M82 is also known as the Cigar Galaxy.

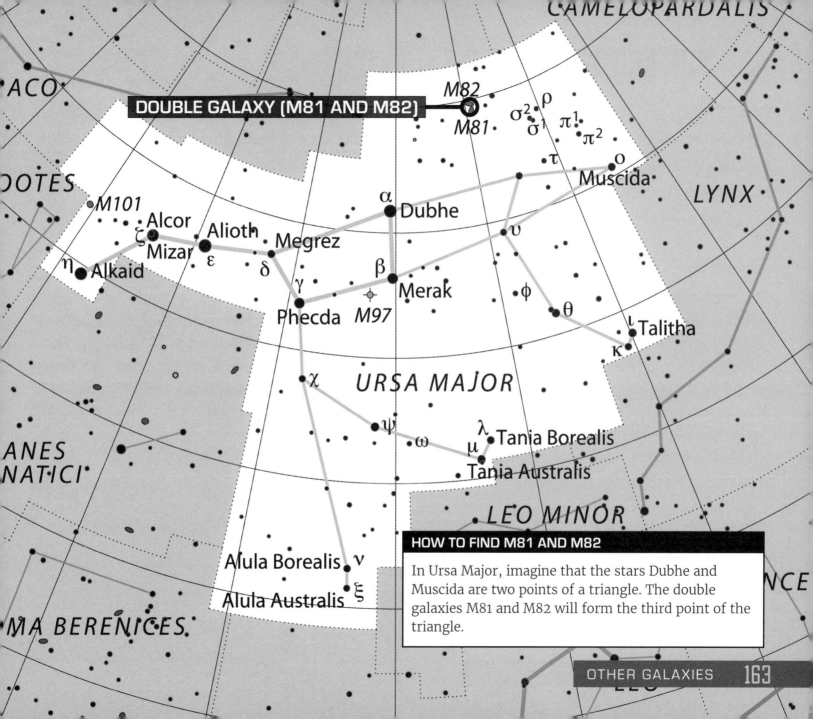

CAMELOPARDALIS

M82

DOUBLE GALAXY (M81 AND M82)

M81

σ² ρ

σ¹ π¹ π²

τ ο

Muscida

LYNX

DRACO

BOOTES

M101

Alcor

ζ Alioth

Mizar

ε δ

α

Dubhe

Megrez

η Alkaid

γ

β

Merak

υ

φ

θ

ι Talitha

κ

Phecda

M97

URSA MAJOR

χ

ψ

ω

λ Tania Borealis

μ

Tania Australis

LEO MINOR

CANES VENATICI

Alula Borealis ν

ξ

Alula Australis

COMA BERENICES

LEO

HOW TO FIND M81 AND M82

In Ursa Major, imagine that the stars Dubhe and Muscida are two points of a triangle. The double galaxies M81 and M82 will form the third point of the triangle.

62. THE LEO TRIPLET OF GALAXIES

OBJECT TYPE: Three spiral galaxies

CONSTELLATION: Leo

APPARENT MAGNITUDE OF M65: +10.25

APPARENT MAGNITUDE OF M66: +8.9

APPARENT MAGNITUDE OF NGC 3628: +10.2

COORDINATES OF M65: 11h 18m 55.9s, +13° 05' 32"

COORDINATES OF M66: 11h 20m 15.0s, +12° 59' 30"

COORDINATES OF NGC 3628: 11h 20m 17.0s, +13° 35' 23"

SEASON: March through July; best in May

DIFFICULTY: Medium

This triplet of galaxies is truly a treat. Depending on your telescope's field of view and what eyepiece you use, you should be able to see all three of these galaxies in a single eyepiece viewing. If not, you should be able to easily hop between the three with very little moving of your telescope. M65 and M66 are considered a double galaxy, so at the minimum you should be able to view both in the same field of view without moving the telescope.

These three galaxies form a group where each galaxy has a gravitational effect on the other two. The group as a whole is about 35 million light years away from Earth. The M65/M66 pair are about 160,000 light years apart, while NGC 3628 is about 300,000 light years from them.

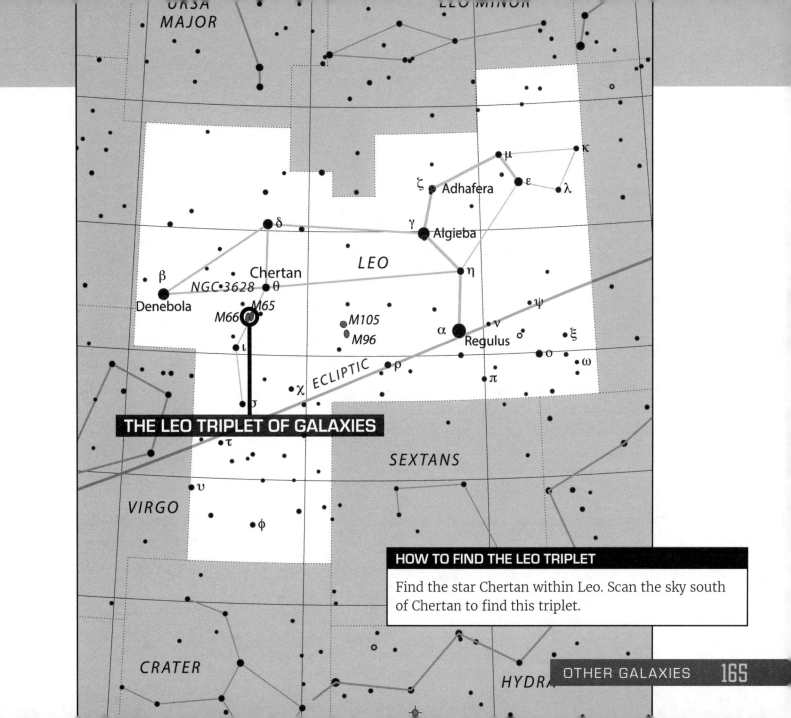

THE LEO TRIPLET OF GALAXIES

HOW TO FIND THE LEO TRIPLET

Find the star Chertan within Leo. Scan the sky south of Chertan to find this triplet.

CHAPTER TEN

THE STARS

Stars have been guiding navigators and travelers since the beginning of time. And they are a nightly reminder of the beauty and variety that the universe holds.

There are about 6,000 stars visible to the naked eye, and it's downright fascinating to see how different and unique they are.

One of the most obvious differences is the varying brightness of the stars—some are extremely bright and beautiful while others are dim and barely visible.

Many stars also radiate brilliant colors other than white. Another variable is a star's intensity, which can modulate over time. Others appear to be a single star but when viewed through a telescope, they are revealed to be two or more stars clustered very close together.

This chapter presents a selection of stars that highlight this diversity.

THE BRIGHTEST AND MOST COLORFUL STARS

It is often thought that Polaris is the brightest star in the sky. This isn't true—Polaris is actually an average, ordinary star, even a bit dim compared to others. However, there are a lot of stars that are so bright that they stand out in terms of visibility and their crispness of color. This section will look at seven of the ten brightest stars in the sky. (The other three in the top ten are so far south that one

[Alpha Centauri] cannot be viewed in North America, while the other two [Canopus and Achernar] are not viewable from most of North America.)

Finding and identifying the brightest stars in the sky has several benefits for you. First, it is quite amazing to look at them through a telescope. They come alive with vibrant light and sometimes with remarkable color. Second, these stars act as guideposts to many of the other objects that can be seen in the night sky.

A good example of this guidepost technique is object 65 on this list. The star Vega is the fifth brightest star in the sky. Being this bright makes it very easy to find, and once you have found it, you will easily find the Ring Nebula (object 45), which is not far from it.

Some stars are quite remarkable celestial objects because they have unusual or vibrant colors. This variance in color is often overlooked by amateur astronomers because we typically think of all stars being a white color. However, as you will see, this is not the case. After you look at the brightest stars, you will turn your telescope to some of the most colorful stars and get a closer look at some of the most stunning objects in the night sky.

THE LIFECYCLE OF A STAR

Two different factors determine why, and when, some stars show varying sizes and colors. The first factor is that stars, just like every other object in the universe, go through a life cycle. And throughout this cycle, their size, temperature, and color changes. The second factor is a star's size and matter composition. How much matter a star starts out with also determines its life cycle and its color changes.

A star starts out as a massive nebulous cloud of gas that, because of gravitational forces, pulls together to form into what we consider to be a star. Nuclear reactions at the center of this formation give off a tremendous amount of light and heat.

An average-sized star, like our sun, will start out as yellow, and over the course of its life will use up its fuel, cool down, and expand into a red dwarf. It will eventually cool further and shed its outer shell, forming a planetary nebula. The core that these stars leave behind is known as a white dwarf.

Larger stars can be a thousand times larger than average yellow stars. These start out hotter and brighter and burn either white or blue. As they use up their fuel, they cool and turn into a red or orange supergiant. These stars end their lifecycles in an explosion as either a nova or supernova.

Not every star follows the same colorful path. In our galaxy, each star is at a different point in this life cycle—some of them are young and yellow or bright white, others are older and red or orange, while some are white dwarves at the end of their life. This complete spectrum of colors and lifespans is available for us to observe.

VARIABLE STARS

We don't typically realize it, but some stars vary in their intensity over time. Some vary by a small, almost unnoticeable amount, and others by a significant amount. In this chapter, you will also look at a few of the most dramatic, and fun, variable stars. One varies in both its brightness and color; that one is called Hind's Crimson Star.

There are two general causes for this variation in brightness. Some vary because of intrinsic properties, such as pulsation or eruption within the star. This characteristic is what we first think of. But, there is another main reason why stars appear to vary in brightness. Some stars don't physically change in brightness; we just perceive them to change because of eclipse events. Some stars are in a system of multiple stars or have enormous planets revolving around them. As they revolve, the planet or the dimmer star will move in front of the star, dimming it significantly. It is the same thing that happens when we have an eclipse here on Earth and the moon moves in front of the sun.

TIPS FOR OBSERVING VARIABLE STARS

Variable stars aren't "blinking" stars. You can't look through your telescope and actually see them change in brightness. Instead, they cycle through changes over a period of time. And this cycle varies a lot depending on the star and the reason for its variability. So, when observing variable stars, you should track them over a long period, ranging from several days for Algol to a full year for Mira.

Observing variable stars is a full hobby unto itself, and there are many amateur astronomers who pursue it actively, carefully tracking the magnitude of many variables.

DOUBLE STARS

Lastly, this chapter will showcase a few double stars, which are an interesting phenomenon. Some are simply optical doubles, meaning they seem to be close together because of our line of sight, but one is actually much farther away than the other. Another type of double star is a binary star. This is where the two stars orbit around each other. They are interesting to look at through a telescope, particularly when they are of different colors.

Did you know that the North Star, Polaris, is a double star? On a good viewing night with a dark sky and a medium-sized telescope, you can see both stars. And in the Ursa Major, there is the famous double star pair called Alcor and Mizar. When you turn your telescope to this double, you discover that Mizar itself is a double within the double. It has a smaller and dimmer star very close to it.

The distance between stars is measured in arc seconds, which is a very small measure of angular distance. The smaller the arc second number, the closer together the stars. As a reference, the moon is about 1800 arc seconds in diameter.

If your telescope comes with a moon filter that screws onto the eyepiece, you can try using it when observing double stars. It often helps in discerning between the two stars.

63. SIRIUS

OBJECT TYPE: Brightest visible star; white main sequence star

CONSTELLATION: Canis Major

APPARENT MAGNITUDE: -1.46

COORDINATES: 06h 45m 08.92s, -16° 42' 58.02"

SEASON: January through April

DIFFICULTY: Easy

Sirius, the Dog Star, is a star of legend. Being the brightest star in the sky means it was important to many of the cultures of humanity throughout our history. Sirius is 8.6 light years away from Earth. It is a main sequence star, which means it is in the stable adult period of its life. A star can stay in this stage of its life for hundreds of millions of years.

In the next two star charts you will see the constellation of Orion, the Mighty Hunter, and the constellation of Canis Major, the Great Dog. Sirius is known as the Dog Star because it is in the constellation of the Great Dog. These two constellations, the hunter and his dog, have been climbing up into the sky every year since the dawn of history.

Have you ever heard the expression "The Dog Days of Summer" and wondered exactly what it meant or where it came from? It comes from the ancient Greeks. In late July, Sirius would make its first appearance in the sky just before sunrise, marking the beginning of the hottest days of summer in Greece.

The ancient Egyptians also used Sirius to predict the flooding of the Nile River. Every winter, Sirius would spend seventy days unseen below the horizon, and its first appearance in the sky denoted that the flooding of the Nile would soon occur.

Nowadays we don't have any beliefs about the annual appearance of Sirius, but we nonetheless look forward to its appearance because of its brilliance and beauty.

Sirius is also a great marker star that will help you identify many other dimmer objects in the sky.

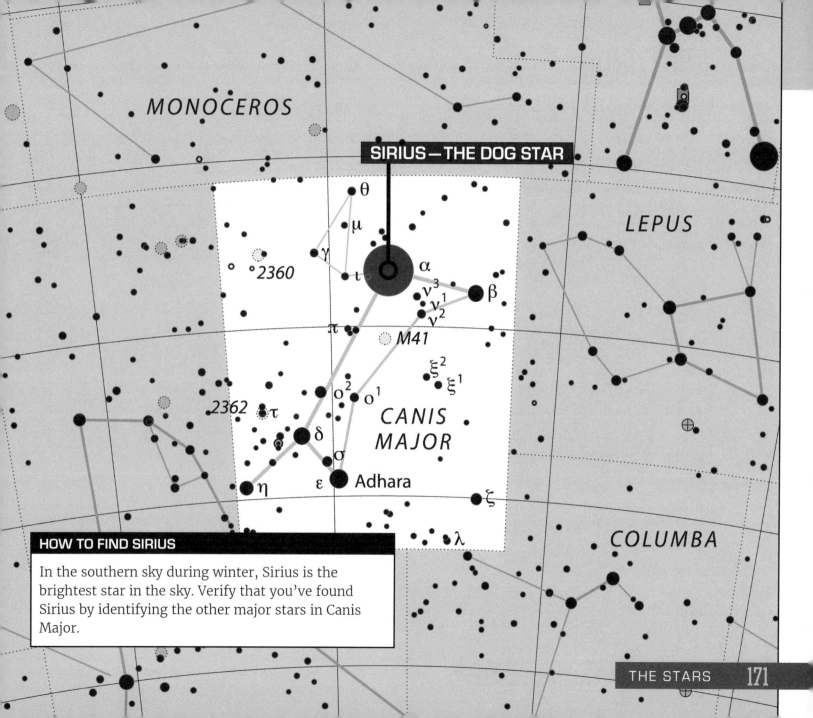

MONOCEROS

LEPUS

SIRIUS — THE DOG STAR

θ

μ

γ

ι

○ 2360

α

ν³
ν¹
ν²

β

π

M41

ξ²
ξ¹

o² o¹

2362 τ

CANIS
MAJOR

δ

σ

η   ε Adhara

ζ

λ

COLUMBA

HOW TO FIND SIRIUS

In the southern sky during winter, Sirius is the brightest star in the sky. Verify that you've found Sirius by identifying the other major stars in Canis Major.

64. ARCTURUS

OBJECT TYPE: Fourth brightest visible star; red giant

CONSTELLATION: Boötes

APPARENT MAGNITUDE: -0.05

COORDINATES: 14h 15m 39.7s, +19° 10' 56"

SEASON: Available year-round in most of North America

DIFFICULTY: Easy

Arcturus is also called the Bear Star. In Greek, Arcturus means "Guardian of the Bear." This is because it is close to Ursa Major, the Large Bear.

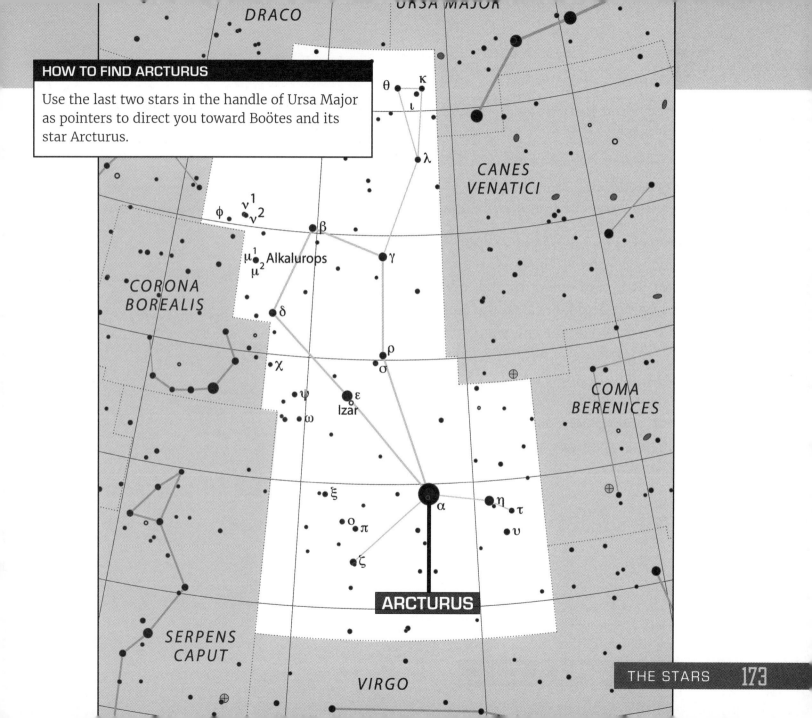

DRACO

URSA MAJOR

HOW TO FIND ARCTURUS

Use the last two stars in the handle of Ursa Major as pointers to direct you toward Boötes and its star Arcturus.

θ κ
ι
λ

CANES VENATICI

ν¹
ν²
φ
β
μ¹ Alkalurops
μ²
γ

CORONA BOREALIS

δ

ρ
σ

χ

COMA BERENICES

ψ ε
ω Izar

α η τ
ξ υ
o π

ζ

ARCTURUS

SERPENS CAPUT

VIRGO

65. VEGA

OBJECT TYPE: Fifth brightest visible star; blue-tinged white main sequence star

CONSTELLATION: Lyra

APPARENT MAGNITUDE: +0.026

COORDINATES: 18h 36m 56.34s, +38° 47' 01.28"

SEASON: April through October; best in summer

DIFFICULTY: Easy

The constellation of Lyra is the Harp, so Vega, as the brightest star in the constellation, is also referred to as the Harp Star. It is also one of the three bright stars that make up the asterism called the Summer Triangle. Vega is a very vibrant and energetic star with a surface temperature of 17,000°F, which is twice that of our sun. It rotates at the extremely fast rate of one revolution every 12.5 hours; in comparison, our sun rotates once every twenty-eight days. This rapid rotation causes Vega to bulge considerably at its equator.

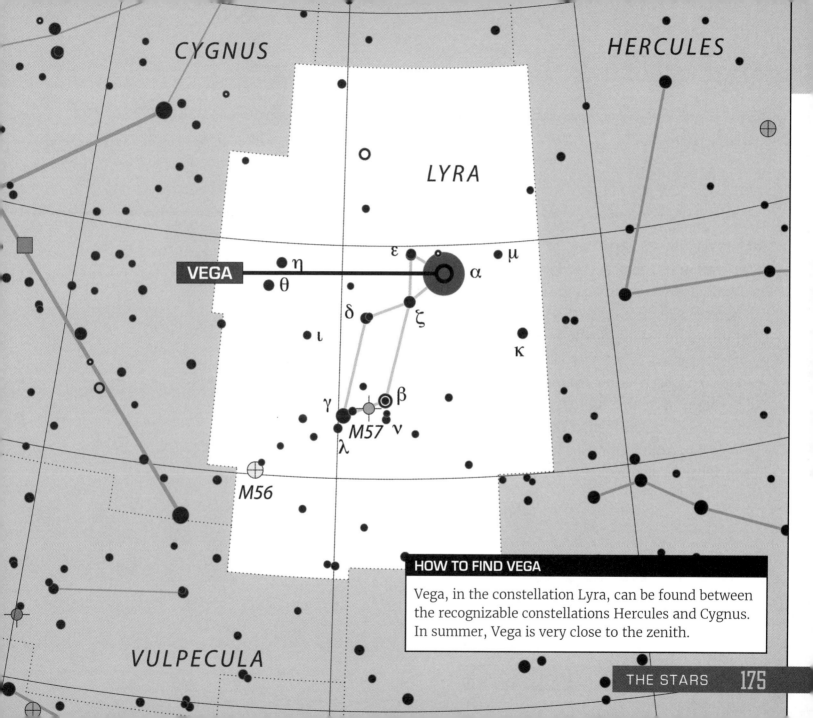

CYGNUS

HERCULES

LYRA

η
θ
VEGA
ε
α
μ

δ
ζ
ι
κ

γ
β
M57
ν
λ

M56

VULPECULA

HOW TO FIND VEGA

Vega, in the constellation Lyra, can be found between the recognizable constellations Hercules and Cygnus. In summer, Vega is very close to the zenith.

66. CAPELLA

OBJECT TYPE: Sixth brightest visible star; yellow giant

CONSTELLATION: Auriga

APPARENT MAGNITUDE: +0.08

COORDINATES: 05h 16m 41.36s, +45° 59' 52.77"

SEASON: December through April; best in February and early March

DIFFICULTY: Easy

Capella was always been thought to be a single yellow giant star, and it appears as such with a telescope. But with powerful astronomical instruments that can see wavelengths other than visible light, it has been discovered that Capella is actually two yellow giant stars in close orbit around each other. They are closer to each other than Earth is to the sun. This system of double stars is even more complex in that there is a nearby pair of red dwarf stars that are locked in orbit around each other, and as a pair, they orbit the Capella double yellow giants.

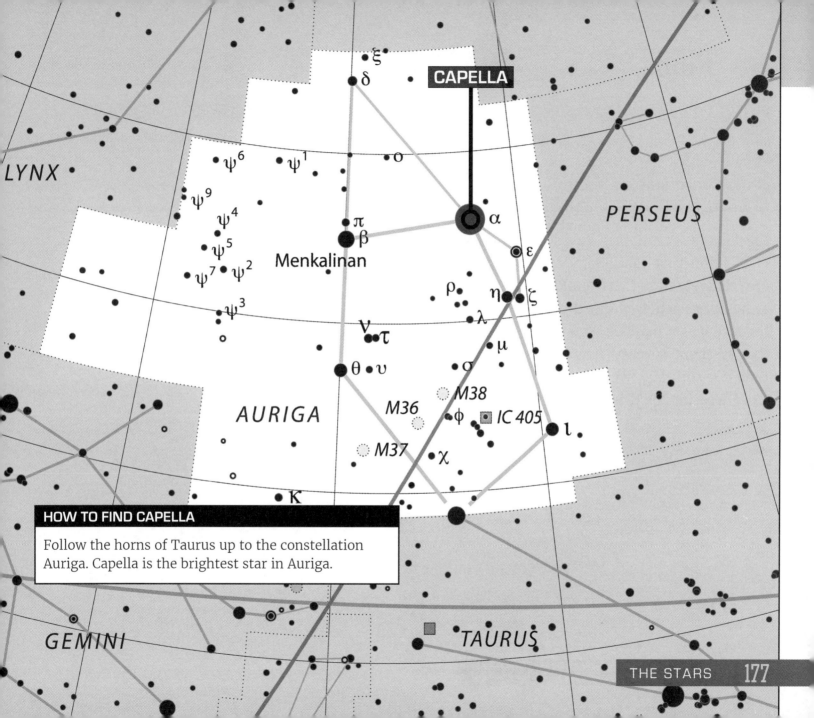

CAPELLA

ξ
δ

LYNX

ψ⁶   ψ¹   ο

ψ⁹

ψ⁴   π   α
β   ε
ψ⁵   **Menkalinan**
ψ⁷ ψ²   ρ   η   ζ
ψ³   λ
                    ν τ   μ
              θ • υ   σ

AURIGA   M38
                M36   φ   ■ IC 405
                M37   χ   ι

                                PERSEUS

κ

HOW TO FIND CAPELLA

Follow the horns of Taurus up to the constellation
Auriga. Capella is the brightest star in Auriga.

GEMINI   **TAURUS**

67. RIGEL

OBJECT TYPE: Seventh brightest visible star; blue-white supergiant

CONSTELLATION: Orion

APPARENT MAGNITUDE: +0.13

COORDINATES: 05h 14m 32.27s, +08° 12' 05.91"

SEASON: November through March; best in January

DIFFICULTY: Easy

Rigel is the seventh brightest star in the sky, which is an impressive feat. Many of the brightest stars are relatively close to us, but Rigel is 860 light years away. It has achieved this distinction of brightness because it is a massive blue-white supergiant star that is about seventy-five times larger in radius than our sun, and about 40,000 times brighter! It truly deserves the title of "supergiant."

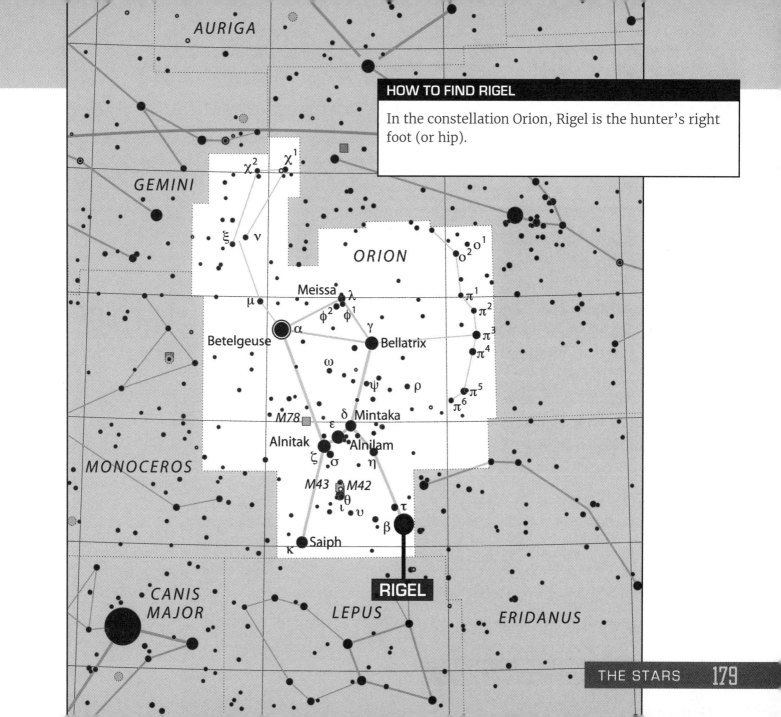

In the constellation Orion, Rigel is the hunter's right foot (or hip).

AURIGA

GEMINI

ORION

χ² χ¹

ξ   ν

Meissa   λ
μ   φ² φ¹
Betelgeuse   α   γ   Bellatrix
                ω
                ψ   ρ

o² o¹

π¹
π²
π³
π⁴
π⁵
π⁶

MONOCEROS

M78   δ Mintaka
      ε
Alnitak   Alnilam
    ζ σ   η
M43   M42
      θ
      ι υ   τ
           β
κ   Saiph

CANIS
MAJOR

LEPUS

ERIDANUS

RIGEL

68. BETELGEUSE

OBJECT TYPE: Tenth brightest visible star; red supergiant

CONSTELLATION: Orion

APPARENT MAGNITUDE: -4.05

COORDINATES: 05h 55m 10.31s, +07° 24' 25.43"

SEASON: November through March; best in January

DIFFICULTY: Easy

Betelgeuse is a fascinating star with a couple of unique characteristics. It is a red supergiant star, which means that it is very big and burns its nuclear fuel at a very fast rate. The star is about 8.5 million years old, which, for a red giant, means it is in the final stages of its life. Scientists estimate that within about 100,000 years, it will erupt into a supernova. Additionally, Betelgeuse is an unusual star because it is a variable star, meaning that it varies in brightness on an irregular schedule. Over the course of a decade, it can vary by as much as 1 full magnitude.

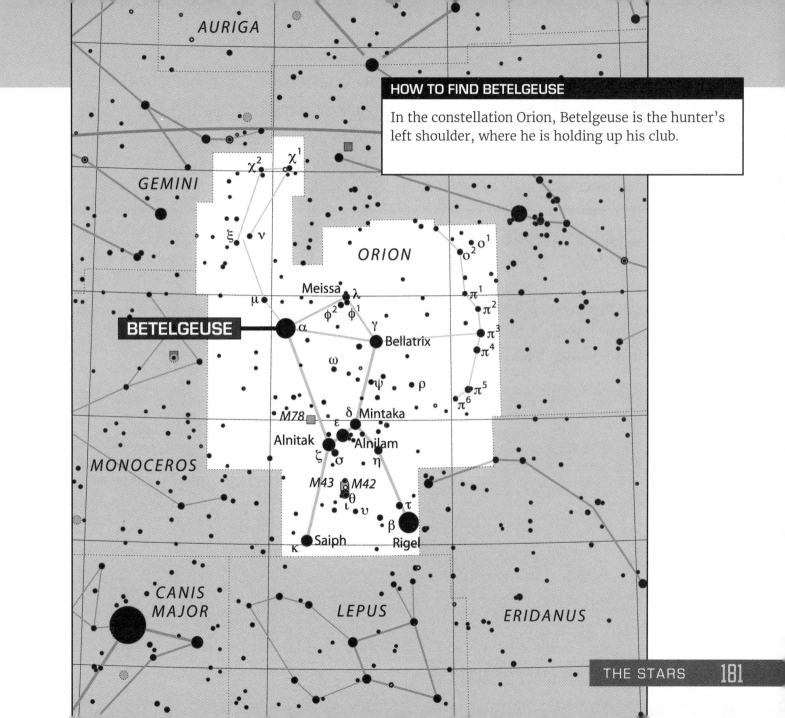

HOW TO FIND BETELGEUSE

In the constellation Orion, Betelgeuse is the hunter's left shoulder, where he is holding up his club.

69. PROCYON

OBJECT TYPE: Eighth brightest visible star; white main sequence star

CONSTELLATION: Canis Minor

APPARENT MAGNITUDE: +0.34

COORDINATES: 07h 39m 18.1s, +05° 13' 29"

SEASON: November through March; best in January

DIFFICULTY: Easy

Sirius is the brightest star in Canis Major, and Procyon, in Canis Minor, is the little brother to Sirius. It is often referred to as the Little Dog Star. To the naked eye, Procyon is one star. Looking at it through a telescope reveals it to be a double star. The second star is a white dwarf with a magnitude of 10.7.

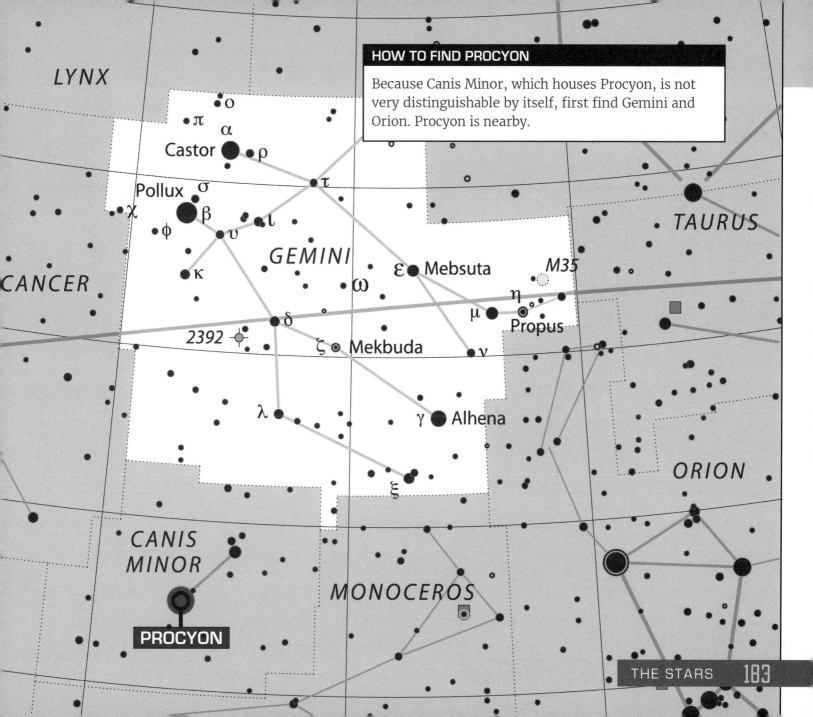

LYNX

ο
π
α
Castor   ρ

Pollux   σ
τ
χ   **β**   ι
φ   υ
GEMINI
κ
CANCER
ω   ε   Mebsuta   *M35*
η
μ   ●
Propus
δ
2392   ζ ◉ Mekbuda   ν
TAURUS

HOW TO FIND PROCYON

Because Canis Minor, which houses Procyon, is not very distinguishable by itself, first find Gemini and Orion. Procyon is nearby.

λ
γ ● Alhena
ξ

ORION

CANIS
MINOR

MONOCEROS

PROCYON

70. BELLATRIX

OBJECT TYPE: Blue supergiant

CONSTELLATION: Orion

APPARENT MAGNITUDE: +1.64

COORDINATES: 05h 25m 07.86s, +06° 20' 58.93"

SEASON: November through March

DIFFICULTY: Easy

Through a telescope, Bellatrix is revealed to have a very deep and rich blue color that borders on being purple. It is also one of the hottest stars in the Milky Way galaxy, at over 38,000°F.

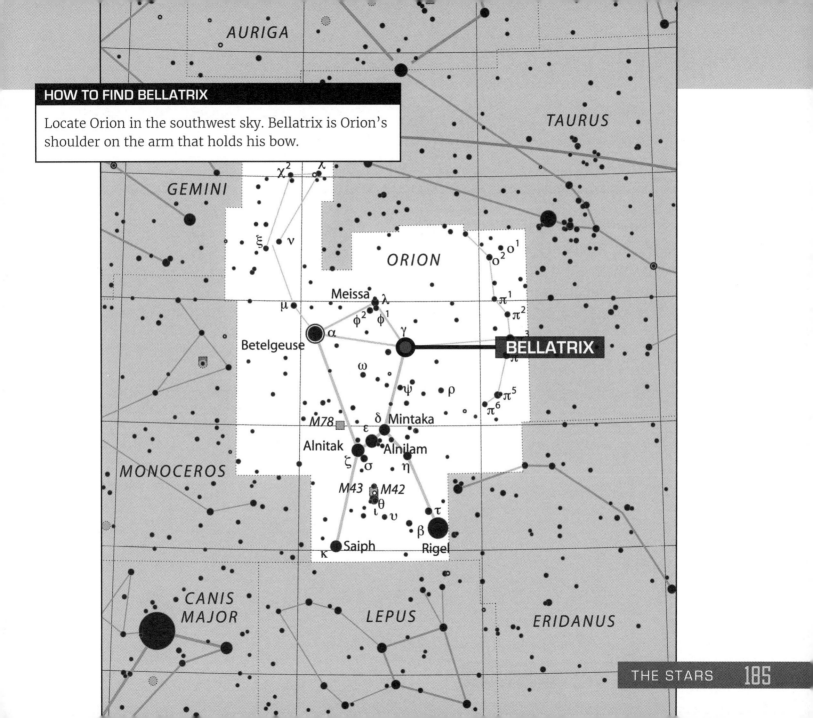

HOW TO FIND BELLATRIX

Locate Orion in the southwest sky. Bellatrix is Orion's shoulder on the arm that holds his bow.

71. ANTARES

OBJECT TYPE: Red supergiant

CONSTELLATION: Scorpius

APPARENT MAGNITUDE: +0.6

COORDINATES: 16h 29m 24.46s, -26° 25' 55.21"

SEASON: June through August; best in July

DIFFICULTY: Easy

This bright red star in Scorpius is often referred to as the heart of the scorpion. A red supergiant, Antares is nearing the end of its life cycle and will eventually explode as a supernova.

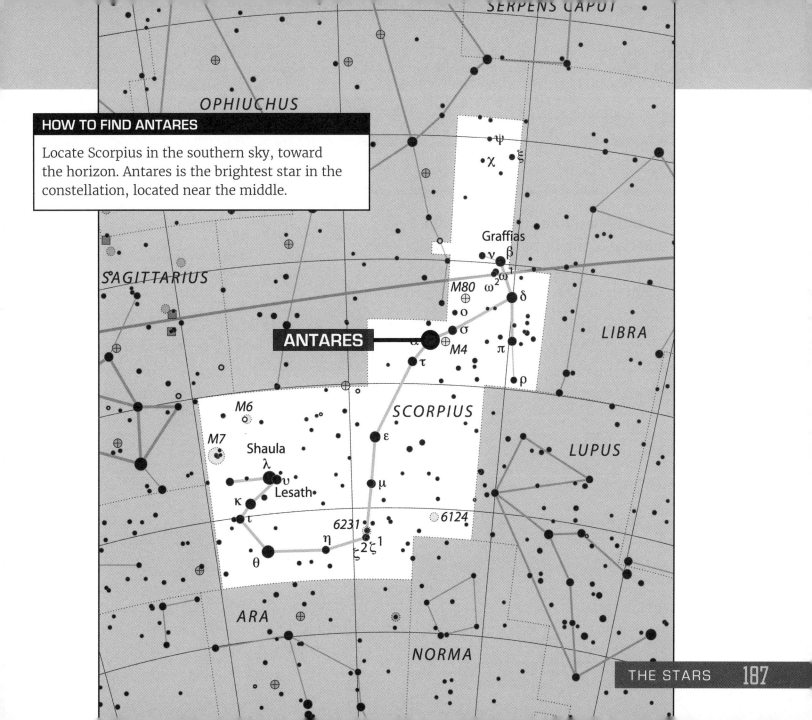

OPHIUCHUS

HOW TO FIND ANTARES

Locate Scorpius in the southern sky, toward
the horizon. Antares is the brightest star in the
constellation, located near the middle.

SAGITTARIUS

Graffias

ν β

ω¹

M80   ω²   δ

o

ANTARES   α   σ

M4

τ   π

ρ

LIBRA

SCORPIUS

M6

M7

ε

Shaula

λ   υ

Lesath   μ

κ

ι

6231   6124

η   θ   ζ2 ζ1

LUPUS

ARA

NORMA

72. ALTAIR

OBJECT TYPE: Main sequence star

CONSTELLATION: Aquila

APPARENT MAGNITUDE: +0.76

COORDINATES: 19h 50m 46.99s, +08° 52' 05.95"

SEASON: July through August; best in August

DIFFICULTY: Easy

Altair, a yellow-white star, is the twelfth brightest star in the sky and one of the three stars that make up the asterism of the Summer Triangle. It is the most prominent star in the constellation Aquila.

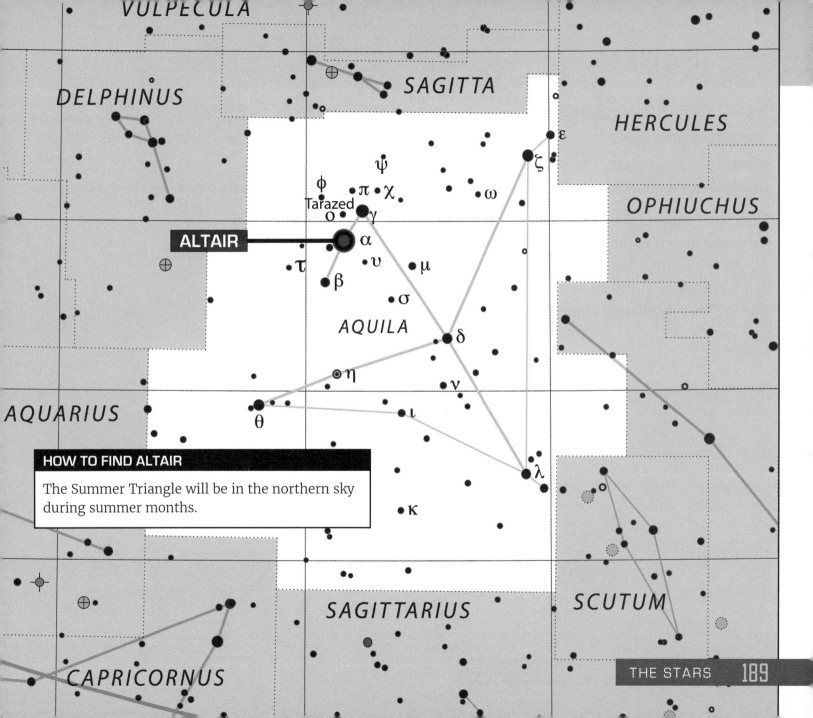

VULPECULA

DELPHINUS

SAGITTA

HERCULES

ψ

φ

π χ

ω

OPHIUCHUS

Tarazed

ο γ

ALTAIR

α

τ

υ μ

β

σ

AQUILA

δ

η

ν

AQUARIUS

θ

ι

λ

HOW TO FIND ALTAIR

The Summer Triangle will be in the northern sky
during summer months.

κ

SCUTUM

SAGITTARIUS

CAPRICORNUS

73. DENEB

OBJECT TYPE: Blue-white supergiant

CONSTELLATION: Cygnus

APPARENT MAGNITUDE: +1.25

COORDINATES: 20h 41m 25.9s, +45° 16' 49"

SEASON: June through December

DIFFICULTY: Easy

This blue-white star is one of the three stars in the asterism known as the Summer Triangle. The name Deneb comes from the Arabic word meaning "hen's tail." The name is fitting, as Deneb sits at the tail-end of the constellation Cygnus, the swan.

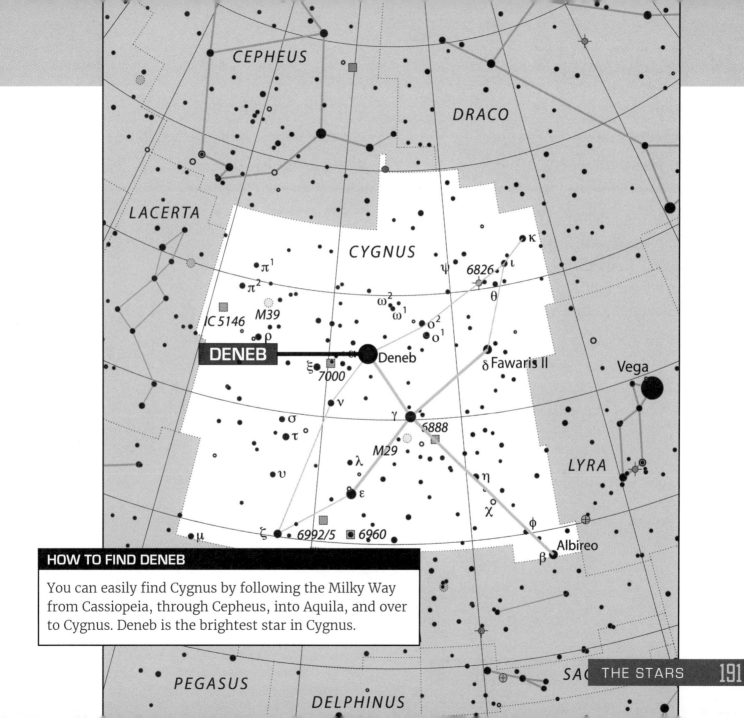

DENEB

HOW TO FIND DENEB

You can easily find Cygnus by following the Milky Way from Cassiopeia, through Cepheus, into Aquila, and over to Cygnus. Deneb is the brightest star in Cygnus.

74. ALDEBARAN

OBJECT TYPE: Orange giant star

CONSTELLATION: Taurus

APPARENT MAGNITUDE: +0.86

COORDINATES: 04h 35m 55.2s, +16° 30' 33"

SEASON: November through March; best in January

DIFFICULTY: Easy

Aldebaran is the brightest star in the constellation of Taurus, the bull, and it is often referred to as the eye of the bull. In medieval times, it was referred to as the heart of the bull. Aldebaran is extremely large and about forty-four times the diameter of our Sun.

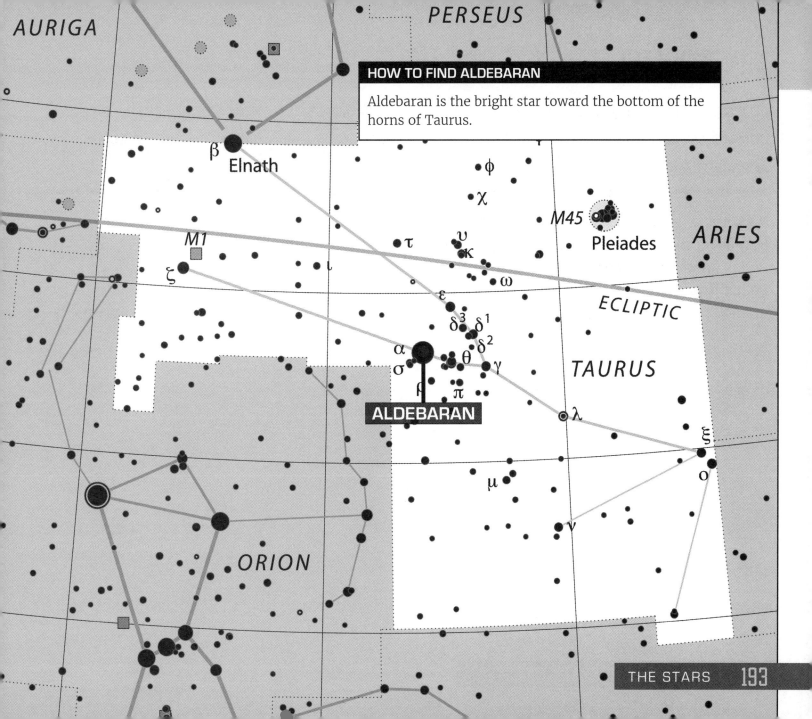

AURIGA

PERSEUS

HOW TO FIND ALDEBARAN

Aldebaran is the bright star toward the bottom of the horns of Taurus.

β Elnath

φ

χ

M45
Pleiades

ARIES

τ

υ
κ

ω

ECLIPTIC

M1

ζ

ι

ε

δ³ δ¹

δ²

α
σ

θ
γ

TAURUS

β

π

ALDEBARAN

λ

ξ

ο

μ

ORION

ν

75. POLLUX

OBJECT TYPE: Main sequence giant star

CONSTELLATION: Gemini

APPARENT MAGNITUDE: +1.14

COORDINATES: 07h 45m 18.95s, +28° 01' 34.32"

SEASON: November through April; best in March

DIFFICULTY: Easy

Pollux is a very close neighbor to Earth, located only about thirty-three light years away. Castor and Pollux are the twin stars in Gemini, but of the two, Pollux is just a little bit brighter. It shines a golden orange, sometimes white.

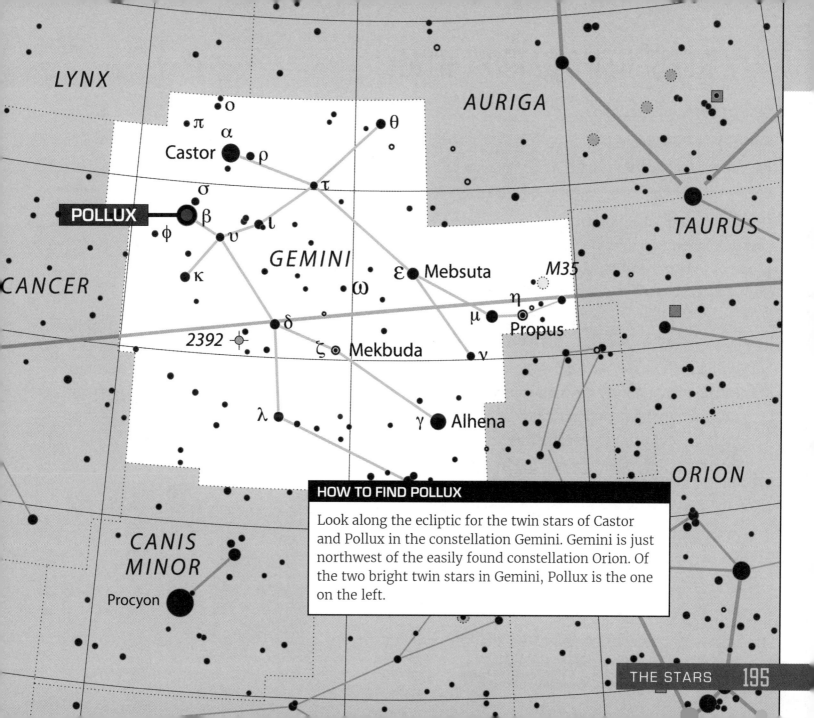

LYNX

AURIGA

ο
π
α
Castor
ρ
θ

σ
τ

POLLUX
β
φ
υ ι
κ

CANCER

GEMINI
ω

ε Mebsuta

M35

η
μ
Propus

δ
2392
ζ Mekbuda
γ

λ
γ Alhena

TAURUS

ORION

CANIS
MINOR

Procyon

HOW TO FIND POLLUX

Look along the ecliptic for the twin stars of Castor
and Pollux in the constellation Gemini. Gemini is just
northwest of the easily found constellation Orion. Of
the two bright twin stars in Gemini, Pollux is the one
on the left.

76. HERSCHEL'S GARNET STAR

OBJECT TYPE: Red giant

CONSTELLATION: Cepheus

APPARENT MAGNITUDE: +4.08

COORDINATES: 21h 43m 30.46s, +58° 46' 48.2"

SEASON: Available year-round for most of North America

DIFFICULTY: Easy

Herschel's star, also known as Mu Cephei, is a red star that is 100,000 times brighter than our sun. This is one of the brightest and biggest stars in the Milky Way galaxy. Herschel discovered this star himself and described it as "a very fine deep garnet color." Since then, it has been referred to as "Herschel's Garnet Star."

WILLIAM HERSCHEL

William Herschel was an eighteenth-century astronomer famous for many advances in the field, including the construction of his famous 40-foot telescope, which was the largest telescope of the time. He was the first president of the Royal Astronomical Society. But his most famous accomplishment is the discovery of Uranus. It was the first planet to be discovered that was not visible to the naked eye.

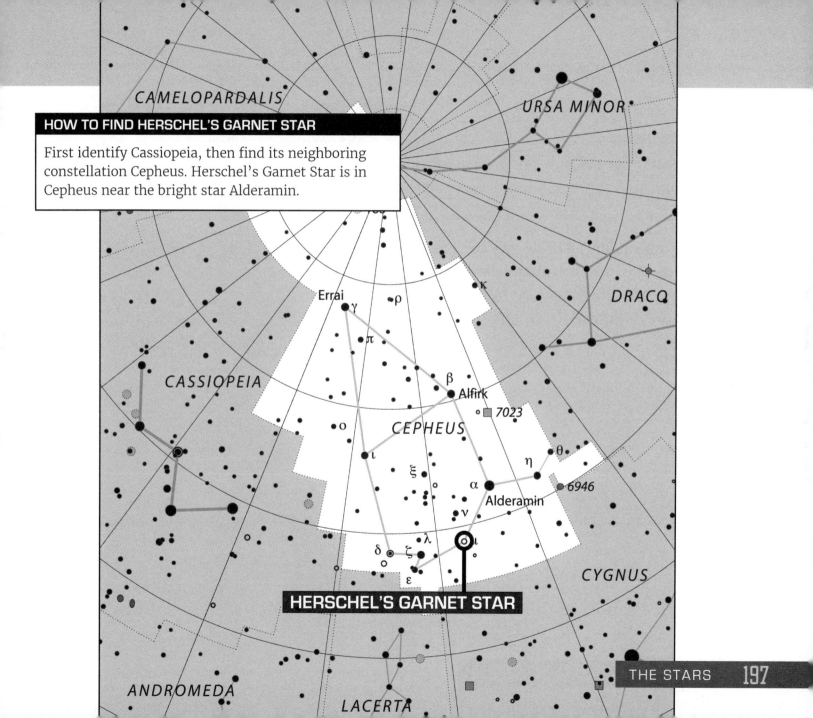

HOW TO FIND HERSCHEL'S GARNET STAR

First identify Cassiopeia, then find its neighboring constellation Cepheus. Herschel's Garnet Star is in Cepheus near the bright star Alderamin.

CAMELOPARDALIS

URSA MINOR

DRACO

Errai
γ
ρ
κ
π
β
CASSIOPEIA
Alfirk
7023
o
CEPHEUS
ι
η
θ
ξ
α
6946
ν
Alderamin
δ
ζ
λ
ι
ε
HERSCHEL'S GARNET STAR
CYGNUS

ANDROMEDA
LACERTA

77. ALGOL

OBJECT TYPE: Eclipsing binary variable star

CONSTELLATION: Perseus

VARIANCE IN APPARENT MAGNITUDE: +3.39 to +2.12

VARIABLE PERIOD: 2.87 days

COORDINATES: 03h 08m 10.13s, +40° 57' 20.33"

SEASON: Available year-round for most of North America

DIFFICULTY: Easy

Algol is an excellent first variable star for you to observe because it is very bright and easily found with the naked eye. It changes by over one full magnitude within a period of just less than three days.

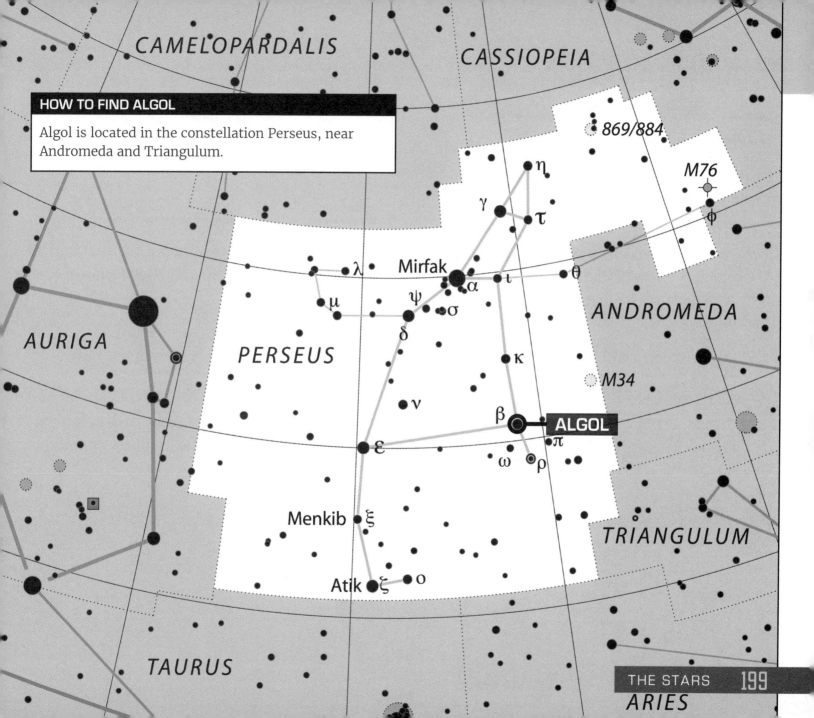

CAMELOPARDALIS

CASSIOPEIA

HOW TO FIND ALGOL

Algol is located in the constellation Perseus, near Andromeda and Triangulum.

869/884

M76

η

γ

τ

φ

λ

Mirfak

θ

α

ι

ANDROMEDA

μ

ψ

σ

AURIGA

δ

PERSEUS

κ

M34

ν

β

ALGOL

ε

π

ω

ρ

Menkib

ξ

TRIANGULUM

Atik

ζ

ο

TAURUS

ARIES

78. DELTA CEPHEI

OBJECT TYPE: Pulsing Cepheid variable star

CONSTELLATION: Cepheus

VARIANCE IN APPARENT MAGNITUDE: +4.37 to +3.48

VARIABLE PERIOD: 5.37 days

COORDINATES: 22h 29m 10.26s, +58° 24' 54.71"

SEASON: Available year-round for most of North America

DIFFICULTY: Easy

The constellation of Cepheus is full of variable stars. It is famous for this. We have given these stars their own classification, calling them Cepheid variables. And of all of them, Delta Cephei is a prototypical example and a great observation for you. It varies by almost a full magnitude in a period just over five days.

VIEWING AND DRAWING TIP

Zeta Cephei is a magnitude 3.35, which is just about the same as our variable Delta Cephei when it is at its peak brightness. Observe the variable over the course of five days and you will be able to discern how it changes by comparing it to Zeta. This is a tried-and-true method for observing variable stars—you compare them to the stars around them.

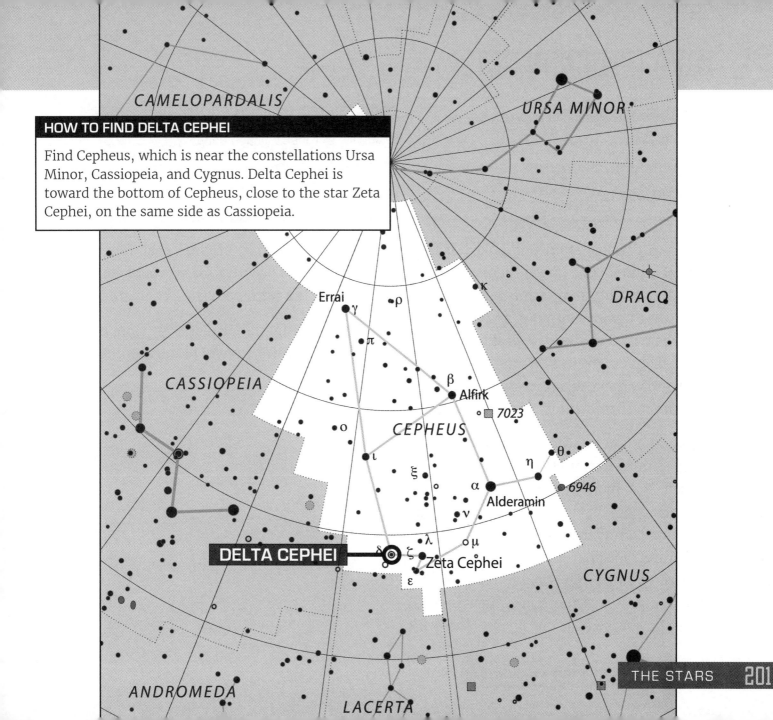

HOW TO FIND DELTA CEPHEI

Find Cepheus, which is near the constellations Ursa Minor, Cassiopeia, and Cygnus. Delta Cephei is toward the bottom of Cepheus, close to the star Zeta Cephei, on the same side as Cassiopeia.

CAMELOPARDALIS

URSA MINOR

DRACO

CASSIOPEIA

Errai
γ

ρ

κ

π

β
Alfirk

7023

CEPHEUS

ο

ι

ξ

η

θ

α

6946

Alderamin

ν

λ

μ

δ
ζ
DELTA CEPHEI
Zeta Cephei

ε

CYGNUS

ANDROMEDA

LACERTA

79. MIRA (OMICRON CETI)

OBJECT TYPE: Pulsating Variable

CONSTELLATION: Cetus

VARIANCE IN APPARENT MAGNITUDE: +10 to +2

VARIABLE PERIOD: 332 days

COORDINATES: 02h 19m 20.79s, -02° 58' 39.49"

SEASON: October through January

DIFFICULTY: Medium

The previous two variable stars you looked at had a variability of around one magnitude, which is noticeable and nice. But, some variable stars, like this one, will vary by an enormous amount. They are visible to the naked eye at their peak yet totally disappear as their brightness wanes. With this dramatic swing in variability, it takes a significantly longer period of time to see the change in magnitude.

The constellation of Cetus is not very bright. There are not a lot of noteworthy stars in it except for Menkar, which is a magnitude 2.5. But, when it comes to variable stars, having a nearby bright star gives you the opportunity to observe the astonishing variability of Mira. If you plan on using your telescope over the course of time, you should observe Mira and make note of its brightness in comparison to Menkar. Over the course of about a year Mira will be as bright as or even brighter than Menkar. Then it will fade into obscurity, being visible only with a telescope. It is a rather amazing characteristic.

Did you know that *mira* is Latin for wonderful? It is an apt name for this marvel of the night sky.

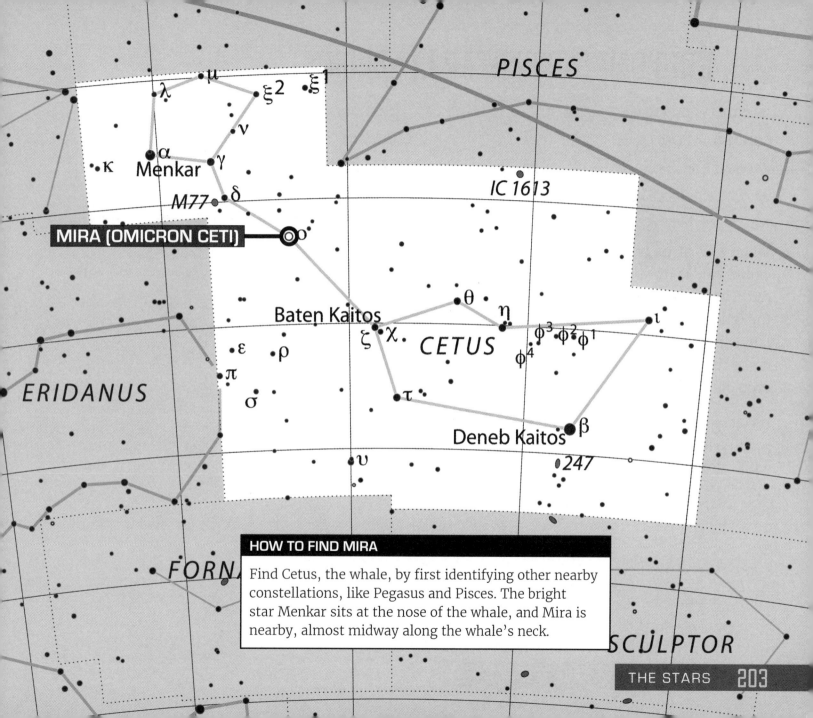

PISCES

μ
λ   ξ2   ξ1
           ν
    α
κ   Menkar   γ
        M77   δ
MIRA (OMICRON CETI)   ο

IC 1613

ERIDANUS

θ
η
Baten Kaitos
ζ   χ   *CETUS*   φ³ φ² φ¹
ε   ρ   φ⁴
π
σ   τ
                              ι

Deneb Kaitos   β

υ   247

FORN   SCULPTOR

HOW TO FIND MIRA

Find Cetus, the whale, by first identifying other nearby
constellations, like Pegasus and Pisces. The bright
star Menkar sits at the nose of the whale, and Mira is
nearby, almost midway along the whale's neck.

80. HIND'S CRIMSON STAR (R LEPORIS)

OBJECT TYPE: Pulsating variable

CONSTELLATION: Lepus

VARIANCE IN APPARENT MAGNITUDE: +11.7 to +5.5

VARIABLE PERIOD: 427 days

COORDINATES: 04h 59m 36.34s, -14° 48' 22.51"

SEASON: January through February

DIFFICULTY: Medium

When it is at its brightest, Hind's Crimson Star is white, and as it wanes to its dimmest, it changes to red. There are a lot of claims about this star being the reddest star in the sky. This can be debated, but J. R. Hind, the nineteenth-century astronomer who made the star famous, described it as being "like a drop of blood on a black field."

If you are interested in variable stars, this one is well worth a look, as it varies significantly in brightness while also varying in color.

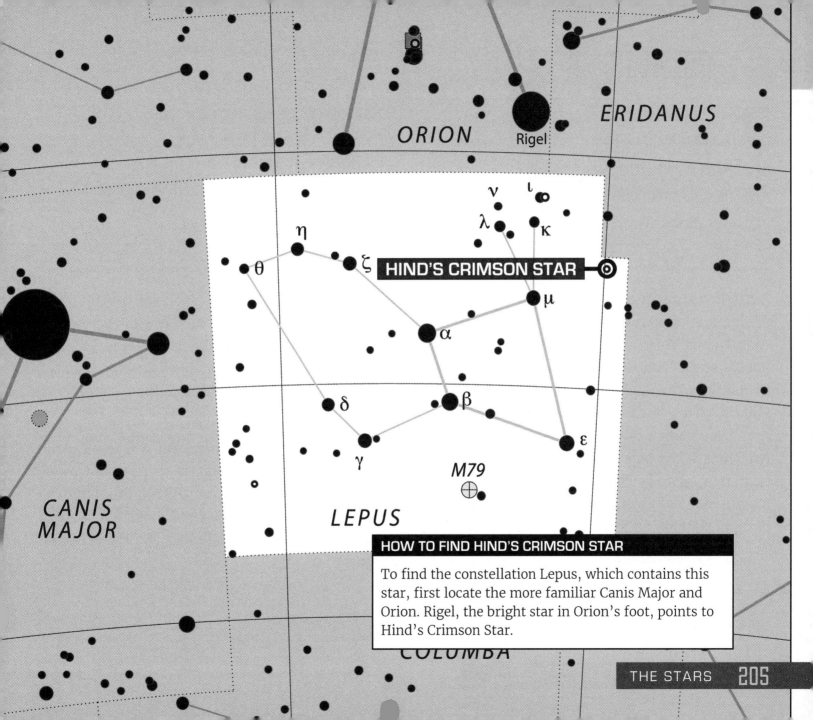

ERIDANUS

ORION
Rigel

ν ι
λ κ

η
θ ζ

HIND'S CRIMSON STAR ⊙

μ

α

CANIS
MAJOR

δ β

γ ε

M79 ⊕

LEPUS

HOW TO FIND HIND'S CRIMSON STAR

To find the constellation Lepus, which contains this
star, first locate the more familiar Canis Major and
Orion. Rigel, the bright star in Orion's foot, points to
Hind's Crimson Star.

COLUMBA

81. POLARIS

OBJECT TYPE: Binary star

CONSTELLATION: Ursa Minor

APPARENT MAGNITUDE OF POLARIS: +2

APPARENT MAGNITUDE OF SECONDARY STAR: +9

COORDINATES OF POLARIS:
02h 31m 48.7s, +89° 15' 51"

COORDINATES OF SECONDARY STAR:
02h 31m 54.15s, +89° 15' 51.30"

SEPARATION: 18 arc seconds

SEASON: Available year-round for most of North America

DIFFICULTY: Easy

Amateur astronomers have a lot of fun with the double star that is Polaris and its much dimmer companion. They call the technique of viewing double stars "splitting a double," and in the case of Polaris, you will often hear the term of "Splitting Polaris." This is the testing of your telescope to see if you can resolve the separation between the two stars. It can be a bit of a challenge because the two stars are very close together and Polaris is a magnitude of 2, which can overpower its much dimmer companion, which has a magnitude of 9.

Give it a try. The North Star is easy to find, and depending on a couple of factors like the viewing conditions and the size of your telescope, you may be able to split this double by seeing it as two stars very close together rather than one slightly elongated star.

It may seem counterintuitive, but you should start with your lowest power eyepiece when trying to split this double. This is because splitting it really isn't about magnification; it is more about the aperture of your telescope, its resolution, and the crispness of image.

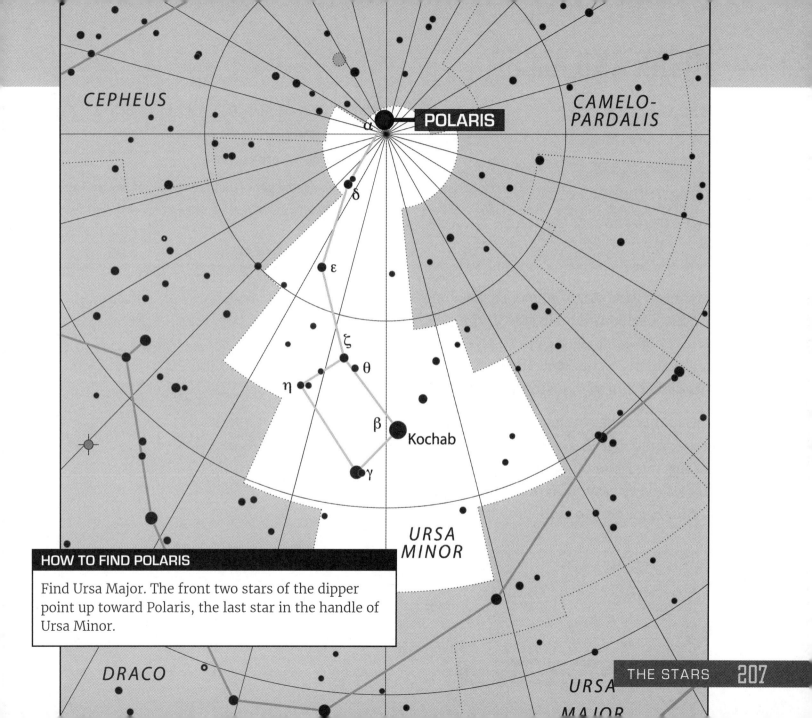

CEPHEUS

CAMELO-
PARDALIS

POLARIS

α

δ

ε

ζ

θ

η

β

Kochab

γ

URSA
MINOR

HOW TO FIND POLARIS

Find Ursa Major. The front two stars of the dipper
point up toward Polaris, the last star in the handle of
Ursa Minor.

DRACO

URSA
MAJOR

82. MIZAR AND ALCOR

OBJECT TYPE: Binary star

CONSTELLATION: Ursa Major

APPARENT MAGNITUDE OF MIZAR: +2.27

APPARENT MAGNITUDE OF ALCOR: +4

COORDINATES OF MIZAR: 13h 23m 55.5s, +54° 55' 31"

COORDINATES OF ALCOR: 13h 25m 13.5s, +54° 59' 17"

SEPARATION: 14 arc seconds

SEASON: Available year-round for most of North America

DIFFICULTY: Easy

Another favorite double star of amateur astronomers is this, the most famous double star in the sky. It has the reputation of being an eye test in ancient times. On clear dark nights, you should be able to resolve this double star with just your naked eye.

Even if you had trouble resolving Polaris, there is no need to worry about this one. It is very easy to resolve in any telescope. And, as it's part of Ursa Major, it is also very easy to locate. This is a great double star for beginners to start out with.

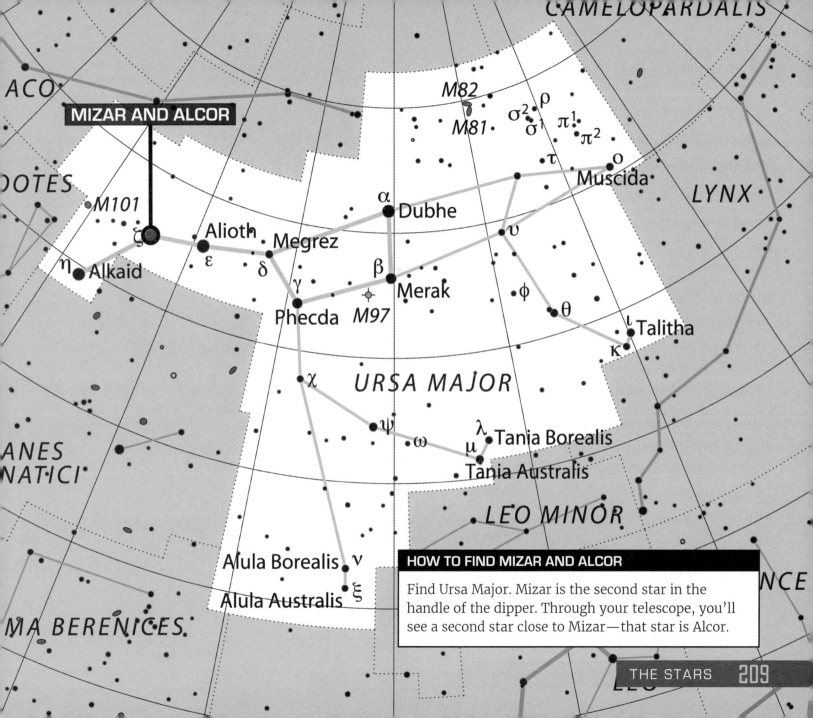

CAMELOPARDALIS

MIZAR AND ALCOR

M82

M81

σ² ρ
σ¹ π¹
π²

τ

ο
Muscida

LYNX

M101

ζ

Alioth

Megrez

α Dubhe

υ

η
Alkaid

ε

δ

β

φ

γ

Merak

θ

Phecda

M97

ι Talitha
κ

χ

URSA MAJOR

ψ

ω

λ Tania Borealis
μ

Tania Australis

LEO MINOR

Alula Borealis ν

ξ

Alula Australis

MA BERENICES

HOW TO FIND MIZAR AND ALCOR

Find Ursa Major. Mizar is the second star in the handle of the dipper. Through your telescope, you'll see a second star close to Mizar—that star is Alcor.

83. CASTOR

OBJECT TYPE: Binary star

CONSTELLATION: Gemini

APPARENT MAGNITUDE OF CASTOR (A): +1.93

APPARENT MAGNITUDE OF CASTOR (B): +2.97

COORDINATES OF CASTOR (A):
07h 34m 36s, +31° 53' 18"

COORDINATES OF CASTOR (B):
07h 34m 36.10s, +31° 53' 18.57"

SEPARATION: 5 arc seconds

SEASON: November through April

DIFFICULTY: Easy

You probably have heard of Castor and Pollux. They are the twin stars in the constellation Gemini. But, did you know that one of these stars, Castor, is a double? In a way, Gemini is a pair of twins that includes another set of twins.

Castor and Pollux are easy to see, but spotting the double star of Castor will take a little bit more work. The two stars that make up this double are very similar in magnitude, and both a brilliant white color. It is a wonderful double to observe and has been referred to as a pair of shining diamonds side by side.

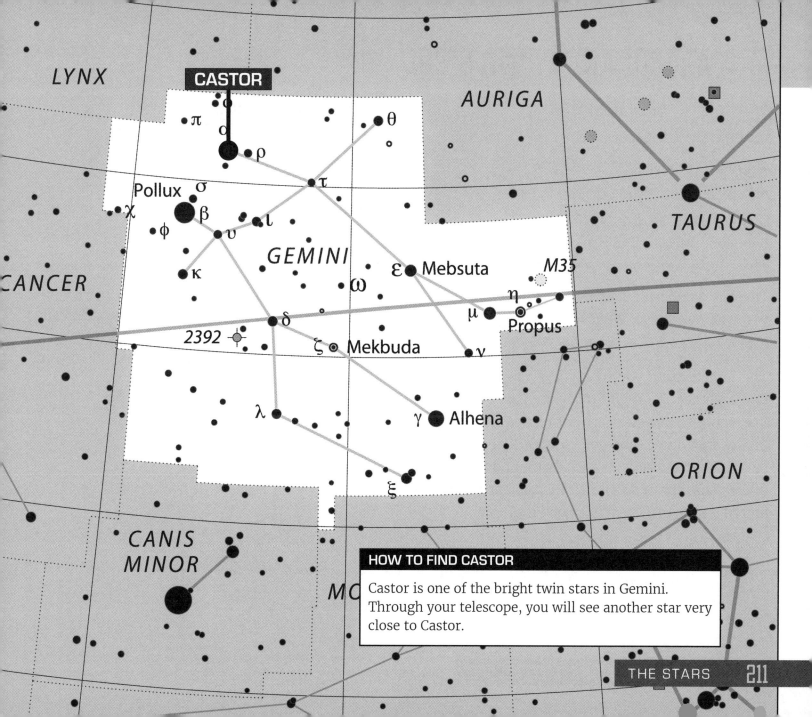

LYNX

CASTOR

AURIGA

π ψ σ

ρ

TAURUS

Pollux σ
χ β
φ υ ι

GEMINI

M35

CANCER

κ

ε Mebsuta

ω

η
μ
Propus

δ

2392

ζ Mekbuda

γ

λ

γ Alhena

ξ

ORION

CANIS
MINOR

MO

HOW TO FIND CASTOR

Castor is one of the bright twin stars in Gemini.
Through your telescope, you will see another star very
close to Castor.

84. ALGEDI AND PRIMA GIEDI

OBJECT TYPE: Optical double star

CONSTELLATION: Capricornus

APPARENT MAGNITUDE OF ALGEDI: +3.57

APPARENT MAGNITUDE OF PRIMA GIEDI: +4.27

COORDINATES OF ALGEDI:
20h 18m 03.25s, -12° 32' 41.47"

COORDINATES OF PRIMA GIEDI:
20h 17m 38.87s, -12° 30' 25.56"

SEPARATION: 6.6 arc minutes

SEASON: September

DIFFICULTY: Easy

This is a wonderful double star that can be resolved with the naked eye. What makes it so lovely is that the two stars are of very similar color and brightness.

The double stars you have looked at so far are all tied together by gravity. They are part of a system. But, some double stars are not related to each other in any way other than coincidence. They appear to be a double, but one is much farther away than the other. This type of double is called an optical double, and Algedi/Prima Giedi is a good example.

The fainter of the two, Algedi, is 570 light years away, and the brighter of the two, Prima Giedi, is 105 light years away. They are both yellow giants and of very similar magnitude.

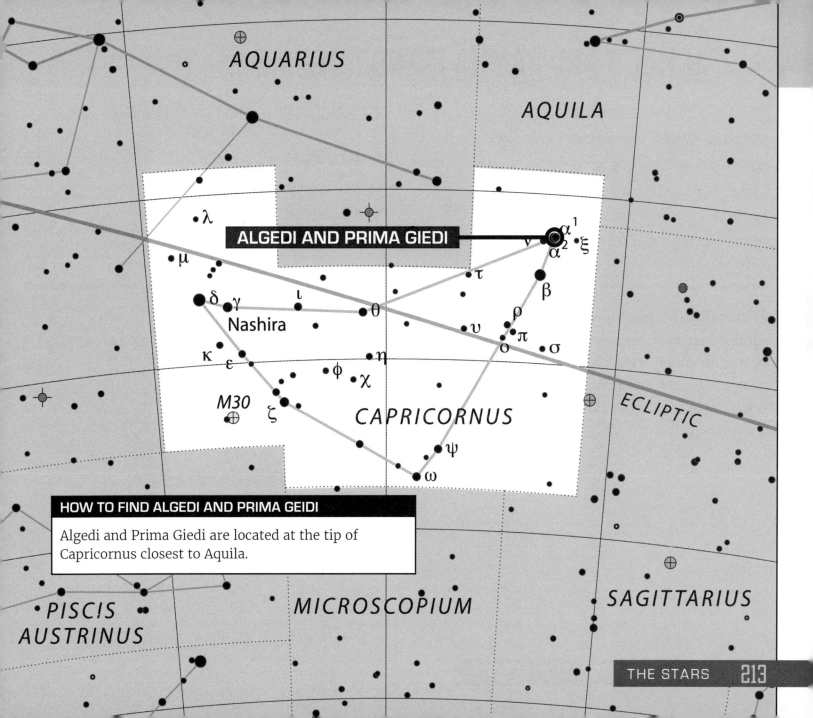

AQUARIUS

AQUILA

λ

ALGEDI AND PRIMA GIEDI

α¹
ν ξ
α²

μ

τ

β

δ γ ι θ
Nashira

ρ
υ π
ο σ

κ ε η
φ χ

M30
ζ ψ
CAPRICORNUS ω

ECLIPTIC

HOW TO FIND ALGEDI AND PRIMA GEIDI

Algedi and Prima Giedi are located at the tip of
Capricornus closest to Aquila.

PISCIS
AUSTRINUS

MICROSCOPIUM

SAGITTARIUS

85. ALMACH AND GAMMA 2 ANDROMEDAE

OBJECT TYPE: Binary double star

CONSTELLATION: Andromeda

APPARENT MAGNITUDE OF ALMACH: +2.26

**APPARENT MAGNITUDE OF
GAMMA 2 ANDROMEDA:** +4.84

COORDINATES OF ALMACH:
02h 03m 53.95s, +42° 19' 47"

COORDINATES OF GAMMA 2 ANDROMEDAE:
02h 03m 54.72s, +42° 19' 51.41"

SEPARATION: 10 arc seconds

SEASON: June through February

DIFFICULTY: Easy

This double, also known as Gamma Andromedae or Almaak, is often considered to be one of the most beautiful double stars because of the contrasting colors of yellow and indigo blue. Well worth a look, it is easy to find and resolve with a small telescope.

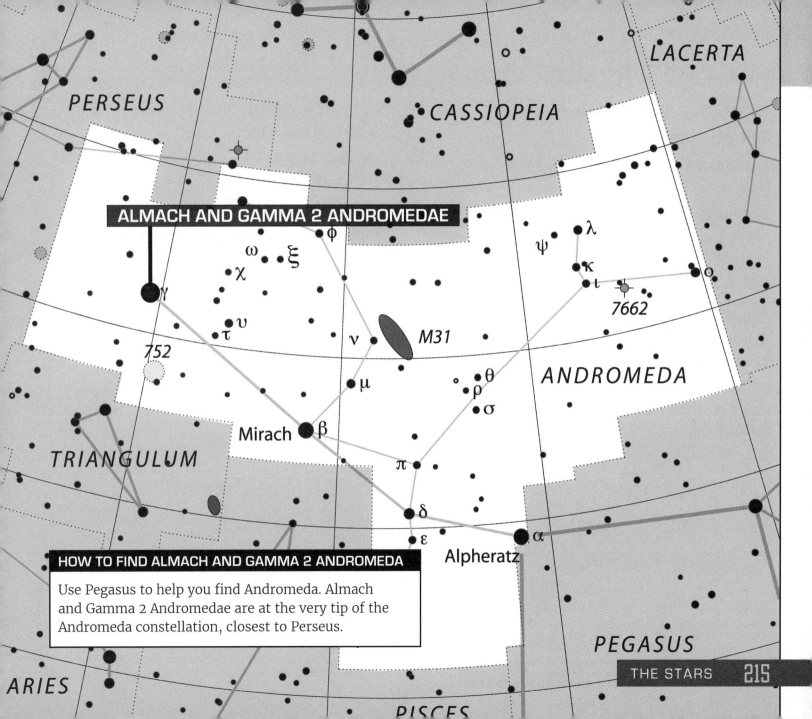

PERSEUS

LACERTA

CASSIOPEIA

ALMACH AND GAMMA 2 ANDROMEDAE

ψ   λ

ω   ξ   κ
χ   φ   ι
γ   θ
752   ν   M31   7662
τ υ   μ   θ
ρ   ANDROMEDA
σ

Mirach   β

TRIANGULUM   π

δ
ε   Alpheratz   α

HOW TO FIND ALMACH AND GAMMA 2 ANDROMEDA

Use Pegasus to help you find Andromeda. Almach
and Gamma 2 Andromedae are at the very tip of the
Andromeda constellation, closest to Perseus.

PEGASUS

ARIES

PISCES

86. RASALGETHI

OBJECT TYPE: Binary double star

CONSTELLATION: Hercules

APPARENT MAGNITUDE OF RASALGETHI (ALPHA 1 HERCULIS): +3.35

APPARENT MAGNITUDE FOR ALPHA 2 HERCULIS: +5.32

COORDINATES FOR RASALGETHI (ALPHA 1 HERCULIS): 17h 14m 38.86s, +14° 23' 25.20"

COORDINATES FOR ALPHA 2 HERCULIS: 17h 14m 39.18s, +14° 23' 23.98"

SEPARATION: 4.8 arc seconds

SEASON: April through November; best in summer

DIFFICULTY: Easy

Alpha 1, which is also known as Rasalgethi, is a red giant star, and Alpha 2 is a yellow giant star. The two, when observed through a telescope, show an excellent contrast in color. The dimmer yellow star will often have a greenish hue because of the contrast against the brighter red star.

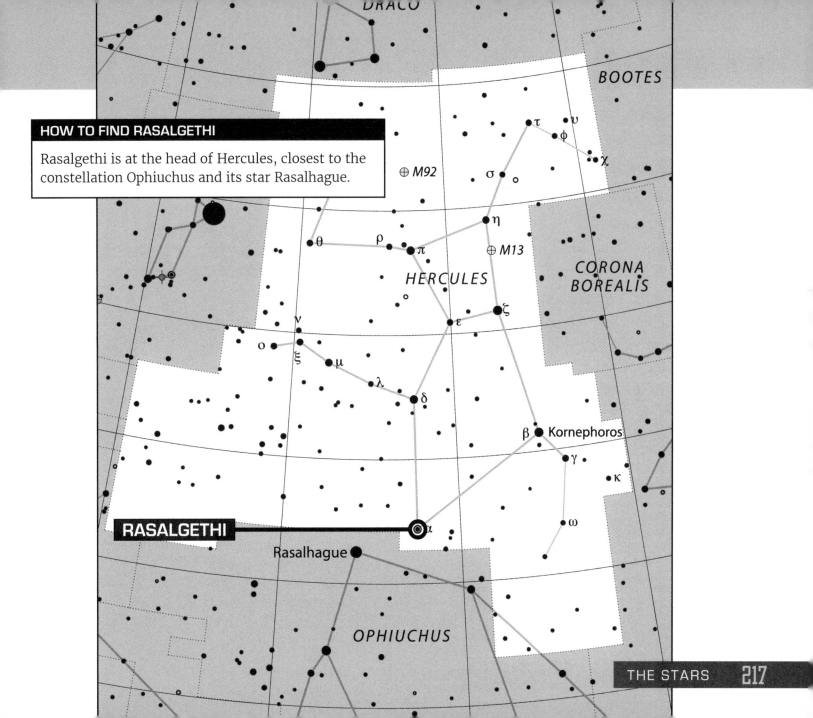

HOW TO FIND RASALGETHI

Rasalgethi is at the head of Hercules, closest to the constellation Ophiuchus and its star Rasalhague.

DRACO

BOOTES

⊕ M92

σ

τ υ
φ
χ

η

⊕ M13

CORONA
BOREALIS

θ ρ
π

HERCULES

ν
ο
ξ
μ
λ
δ

ε
ζ

β ● Kornephoros

γ

κ

ω

RASALGETHI ━━━━━━━━━━━ ◎ α

Rasalhague ●

OPHIUCHUS

87. ALBIREO

OBJECT TYPE: Binary double star

CONSTELLATION: Cygnus

APPARENT MAGNITUDE OF ALBIREO: +3.18

APPARENT MAGNITUDE OF ALBIREO B: +5.09

COORDINATES OF ALBIREO:
19h 30m 43.29s, +27° 57' 34.62"

COORDINATES OF ALBIREO B:
19h 30m 45.39s, +27° 57' 54.99"

SEPARATION: 34.3 arc seconds

SEASON: June through September

DIFFICULTY: Easy

This double star is a favorite among astronomers. The golden-yellow Albireo contrasts very well against the dimmer blue of Albireo B. It is very rewarding to look at the double with the naked eye and only see one star, and then use your small telescope to resolve two different stars of different colors.

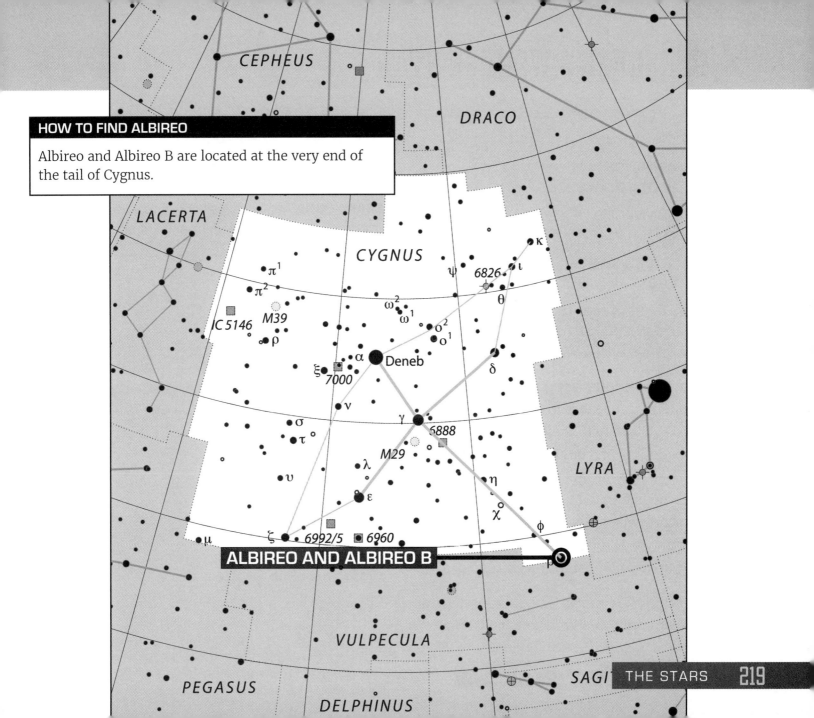

HOW TO FIND ALBIREO

Albireo and Albireo B are located at the very end of the tail of Cygnus.

CEPHEUS

DRACO

LACERTA

CYGNUS

κ

π¹

π²

ψ 6826 ι

ω² θ

ω¹

M39

IC 5146

o²

ρ

o¹

α Deneb

δ

ξ

7000

ν

γ

σ 6888

τ

M29

LYRA

λ

υ

η

ε

χ

ζ 6992/5 6960 φ

μ

ALBIREO AND ALBIREO B β

VULPECULA

PEGASUS

SAGIT

DELPHINUS

88. HERSCHEL'S WONDER STAR

OBJECT TYPE: Triple star system

CONSTELLATION: Monoceros

APPARENT MAGNITUDE OF BETA MONOCEROTIS A: +4.6

APPARENT MAGNITUDE OF BETA MONOCEROTIS B: +5.4

APPARENT MAGNITUDE OF BETA MONOCEROTIS C: +5.6

COORDINATES OF BETA MONOCEROTIS A:
06h 28m 49.07s, -07° 01' 59.03"

COORDINATES OF BETA MONOCEROTIS B:
06h 28m 49.42s, -07° 02' 03.88"

COORDINATES OF BETA MONOCEROTIS C:
06h 28m 49.61s, -07° 02' 04.76"

SEPARATION: The distances between the three stars varies between 2.8 arc seconds and 7.4 arc seconds.

SEASON: November through March

DIFFICULTY: Easy

The stars in this triplet are referred to as Beta Monocerotis. The three stars in the system are Beta Monocerotis A, B, and C. They each appear blue-white, but depending on your telescope and viewing conditions, they may all appear blue or all appear white. The famous eighteenth-century astronomer William Herschel called this triplet "one of the most beautiful sights in the heavens." The three stars together add up to an apparent magnitude of 3.74.

Monoceros is the lesser known Unicorn constellation, but even seemingly insignificant constellations like this hold secret little treasures. The triple star appears to the naked eye to be one bright star, but once you turn your telescope to it, it is revealed as a triplet of stars.

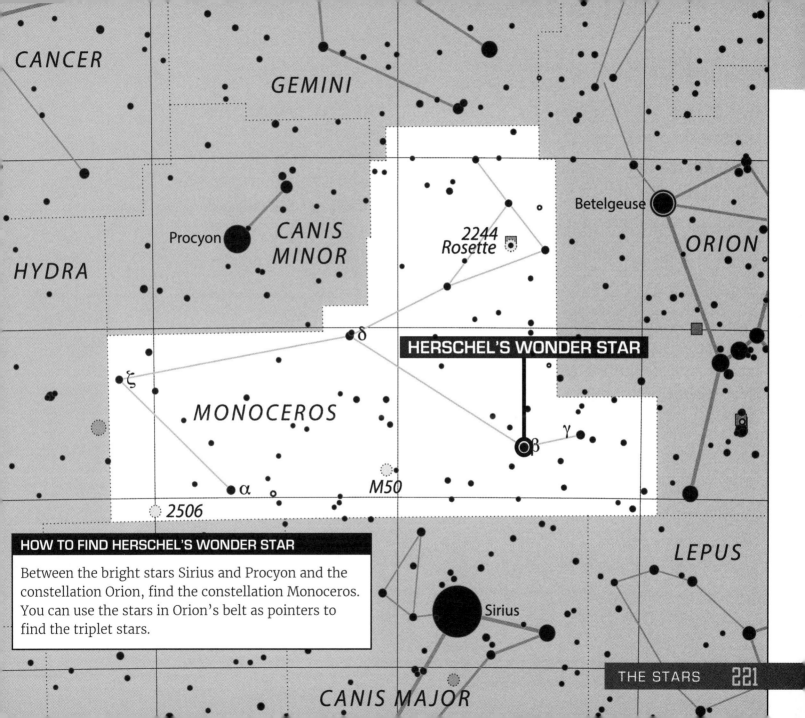

CANCER

GEMINI

HYDRA

Procyon

CANIS
MINOR

2244
Rosette

Betelgeuse

ORION

δ

ζ

HERSCHEL'S WONDER STAR

MONOCEROS

γ

β

α

M50

2506

LEPUS

HOW TO FIND HERSCHEL'S WONDER STAR

Between the bright stars Sirius and Procyon and the
constellation Orion, find the constellation Monoceros.
You can use the stars in Orion's belt as pointers to
find the triplet stars.

Sirius

CANIS MAJOR

89. OMICRON CYGNI

OBJECT TYPE: Optical triple star

CONSTELLATION: Cygnus

APPARENT MAGNITUDE OF 30 CYGNI: +4.83

APPARENT MAGNITUDE OF 31 CYGNI A: +3.80

APPARENT MAGNITUDE OF 31 CYGNI B: +7.01

COORDINATES OF 30 CYGNI:
20h 13m 18.05s, +46° 48' 56.44"

COORDINATES OF 31 CYGNI A:
20h 13m 37.91s, +46° 44' 28.78"

COORDINATES OF 31 CYGNI B:
20h 13m 39.20s, +46° 42' 42.71"

SEASON AVAILABLE: June through December

DIFFICULTY: Medium

This is a beautiful triplet of stars, each with a different color. The pair of 31 Cygni A and B are respectively orange and blue. And they are very close together. The brighter third star, 30 Cygni, is a white star located a small distance away from that pair.

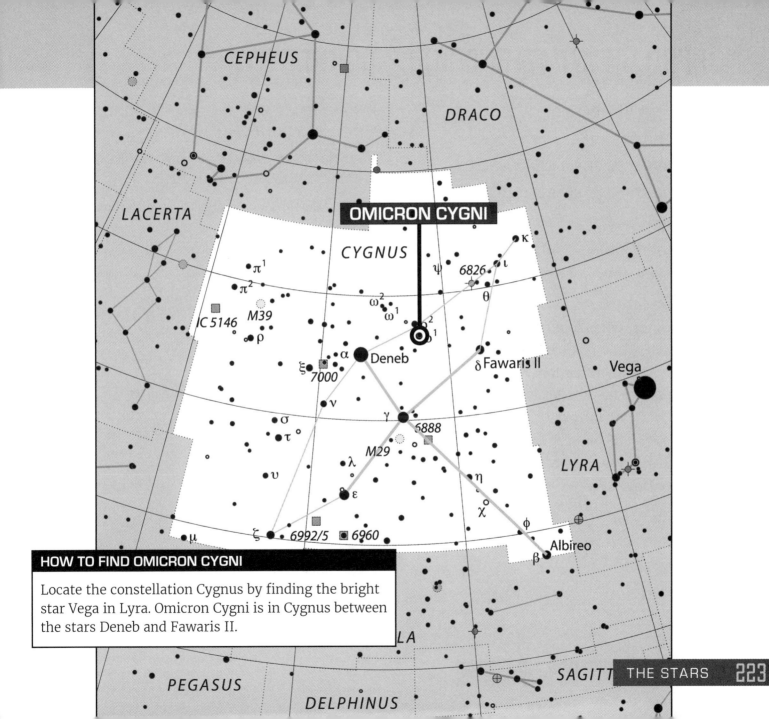

OMICRON CYGNI

CEPHEUS

DRACO

LACERTA

CYGNUS

κ

ψ   6826   ι

θ

π¹

π²

ω²

ω¹

M39

IC 5146

ρ

α   Deneb

ξ

7000

ν

σ

τ

υ

λ   M29

ε

μ   ζ   6992/5   6960

δ Fawaris II

γ   6888

η

χ

φ

Albireo

β

Vega

LYRA

PEGASUS

DELPHINUS

LA

SAGITT

HOW TO FIND OMICRON CYGNI

Locate the constellation Cygnus by finding the bright star Vega in Lyra. Omicron Cygni is in Cygnus between the stars Deneb and Fawaris II.

90. EPSILON LYRAE: A DOUBLE-DOUBLE STAR

OBJECT TYPE: Double binary star

CONSTELLATION: Lyra

APPARENT MAGNITUDE OF EPSILON 1A: +5.02

APPARENT MAGNITUDE OF EPSILON 1B: +6.2

APPARENT MAGNITUDE OF EPSILON 2C: +5.14

APPARENT MAGNITUDE OF EPSILON 2D: +5.5

COORDINATES OF EPSILON 1A AND EPSILON 1B:
18h 44m 20.35s, +39° 40' 12.45"

COORDINATES OF EPSILON 2C AND EPSILON 2D:
18h 44m 22.78s, +39° 36' 45.78"

SEASON: May through October; best in summer

DIFFICULTY: Easy

When looked at with the naked eye, Epsilon Lyrae appears as one star. When looked at with low power or with binoculars, it is resolved into a double star. Switch to a higher magnification, and both of those stars are revealed to be doubles, letting you see all four stars. This is an interesting system of stars because both sets are binary, and the two binary systems revolve around one another—a binary made up of binaries.

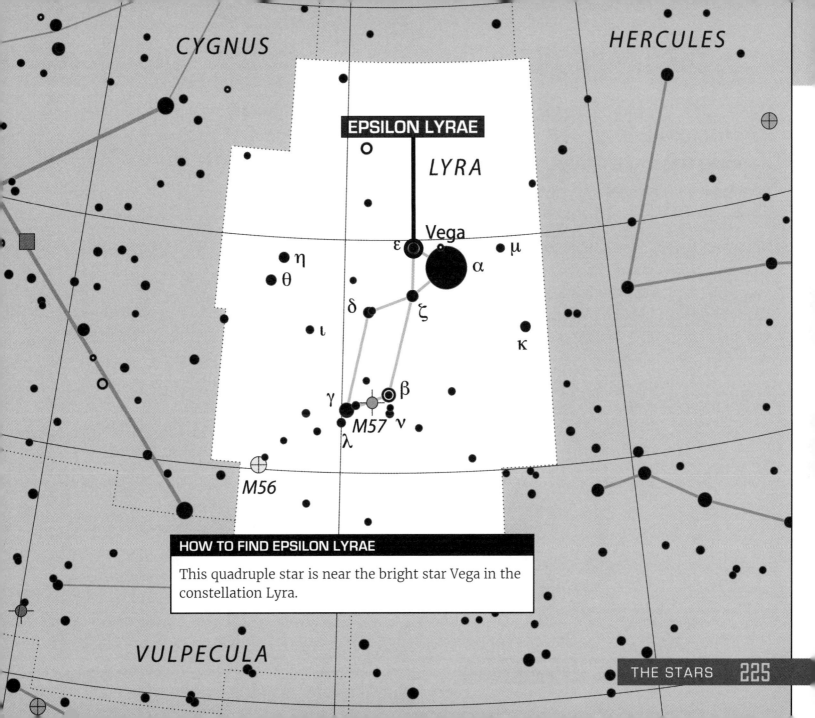

CYGNUS

HERCULES

EPSILON LYRAE

LYRA

Vega

η

θ

ε

α

μ

δ

ζ

ι

κ

γ

β

M57

λ

ν

M56

HOW TO FIND EPSILON LYRAE

This quadruple star is near the bright star Vega in the
constellation Lyra.

VULPECULA

91. THE TRAPEZIUM

OBJECT TYPE: Multiple star system

CONSTELLATION: Orion

APPARENT MAGNITUDE: +4.0

COORDINATES: 05h 35m 17.3s, -05° 23' 28"

SEPARATION: These stars average about 1.5 light years from each other.

SEASON: November through March

DIFFICULTY: Easy

This quadruple star is a double bonus for you because it is embedded in one of the most remarkable nebulae in all the night sky, the M42 Orion Nebula. If you have found that nebula, object 40 on this list, then you have found the Trapezium. But you probably didn't take special note of these stars. Now, you can return to the same location in the constellation and view it in a new way.

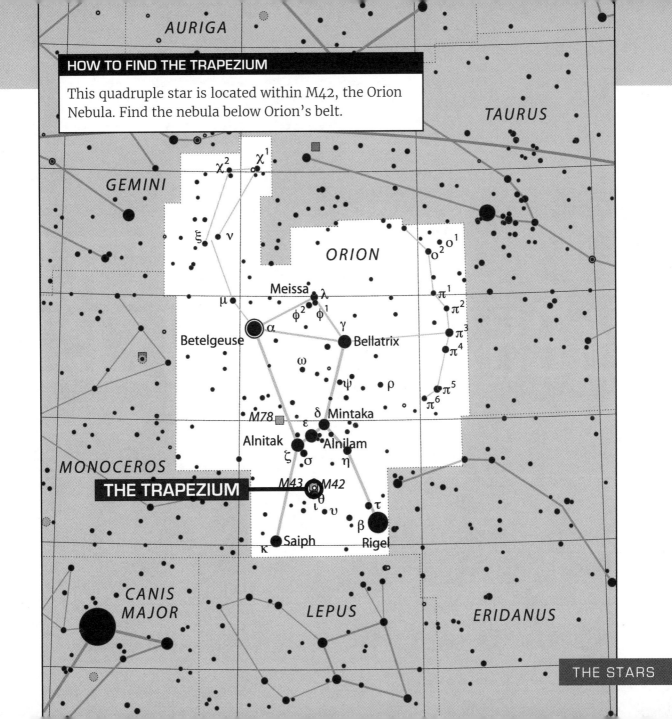

HOW TO FIND THE TRAPEZIUM

This quadruple star is located within M42, the Orion Nebula. Find the nebula below Orion's belt.

CHAPTER ELEVEN

OTHER CELESTIAL OBJECTS

So far you have covered a lot of interesting but common objects in the night sky. Next you will look at some unusual and uncommon things.

The night sky has hundreds of interesting objects, and in this book you have looked at scores of them. But there are other things in the night sky that are either unique unto themselves, or rare to see because they only occur at intervals. In this chapter, you will take a look at some of these objects.

92. METEOR SHOWERS

A meteor shower is a collection of dust and microscopic debris in space. Typically, it is a trail of particles left behind by a comet or meteor. Every year, the Earth passes through these collections, and when these small fragments come in contact with the Earth's atmosphere, it causes friction and they burn up. This creates a falling star.

Because a meteor shower is a field of debris, there can be multiple falling stars while we pass through it. This can be as many as 100 or more per hour, which is why we call them showers. But, the rate of falling stars is unpredictable. It depends on how dense the field of debris is, and we can't accurately predict that. However, we can predict when these showers occur during the year and what part of the sky they originate from.

Generally, you can't watch meteor showers with your telescope. They occur too infrequently and in too wide an area of sky. It is not possible to know where to exactly point your telescope. The best way to view them is with the naked eye. And it can take some time; you may have to watch for a period of an hour or more.

Regrettably, the moon can be a big deterrent to viewing meteor showers because its bright light when full or nearly full can overpower and wash out the dim light from the showers. It can be fruitful to check to see what phase the moon is in before planning an evening observing a meteor shower.

If you want the best opportunity to see as many falling stars as possible, you should try to do your observing on or near the peak date listed.

HOW BEST TO OBSERVE METEOR SHOWERS

Look to the constellation that the current meteor shower is originating from, and give it some time. You may not see a single one for a while, but then you might be surprised by several of them within the span of a minute. Be sure you are under dark skies and your eyes have had at least fifteen minutes in the dark to adjust. If you can stay up late enough, it is often better to view a meteor shower after midnight. This is because the turning of the Earth will spin us right into the debris field.

PROMINENT METEOR SHOWERS THROUGHOUT THE YEAR

METEOR SHOWER	CONSTELLATION	DATES	DESCRIPTION
Quadrantid	Boötes	Dec. 28–Jan. 12 peak: Jan. 4	They originate in the constellation of Boötes and there can be as many as 120 falling stars per hour. This shower is the remnant of an asteroid.
The Lyrids	Lyra	Apr. 21–25 peak: Apr. 22	This meteor shower doesn't have a long open window like other meteor showers. Typically, this is a weak meteor shower with an average of 10 to 20 meteors per hour. But on some years, it unexpectedly peaks at over 100 per hour. You just never know with this shower. It is the remnant of a comet and it has a long history; this shower has been cataloged and noted for over 2,000 years.
Eta Aquarids	Aquarius	Apr. 19–May 29 peak: May 5	This shower originates in the constellation of Aquarius. Typically, it has a rate of about 60 falling stars per hour. This meteor shower occurs as we pass through the remnants of Halley's Comet.
The Arietids	Aries	May 22–Jul. 2 peak: Jun. 7	This is a strong meteor shower but it happens during the day. You can, however, watch for it in the early morning hours before sunrise. Look to the constellation of Aries about an hour before the sun comes up for your best viewing. It can peak at around 60 falling stars per hour.
The South Delta Aquarids	Aquarius	mid-Jul.–mid-Aug. peak: Jul. 28 and Jul. 29	Not as powerful as some of our other meteor showers, the South Delta Aquarids typically has a rate of 15 to 20 falling stars per hour. It is the remnants of two different comets and it can be seen radiating from the constellation of Aquarius.
The Perseids	Perseus	Jul. 29–Aug. 22 peak: Aug. 12	This shower is of special note because it is often spectacular and it is caused by the remnant trail of comet Swift-Tuttle. This meteor shower typically has a rate of 60 meteors per hour.
The Orionids	Orion	Oct. 17–25 peak: Oct. 21 or Oct. 22	This one is a little bit unusual in that it often produces falling stars that are yellow and green. Sometimes it even produces very large and bright meteors. But, its rate is typically rather slow, with an average of 20 meteors per hour.
The Leonids	Leo	Nov. 6–30 peak: Nov. 18	This is not a vibrant meteor shower. It typically has a rate of about 15 meteors per hour.

This is an image of Halley's Comet as it appeared in the sky in 1910.

Comets are regular visitors to our area of the solar system. But most of the time, they are extremely small and difficult to see. For most of them, you need a powerful telescope.

We often think of comets as star-like objects with long tails that streak across the night sky. This is partially true. They do have a bright head and a long tail, but they don't streak across the night sky. To the naked eye they appear stationary which, when they are visible, makes them easy to find. And occasionally, they can get so bright that you can't miss them; they can even get so bright that you can even see them during the day.

Comets travel in elliptical orbits around the sun, and they are rarely in the ecliptic like the planets, so they can be located just about anywhere in the sky at any time. Once an object is identified as a comet it is tracked, meaning its orbit around the sun, and hence its position in the sky, can be predicted with accuracy. However, their brightness is unpredictable, as is the length of their tail. So while a comet is making its way through the Earth's neighborhood, it doesn't mean it will have a comet-like appearance or be very dramatic. Most often, comets appear as fuzzy spots without a tail. The only way to find out how bright it will be, or how dramatic it will look, is to wait for it to get here.

One comet that is showing some promise is 46P/Wirtanen, which will be visible in December 2018. It will peak in brightness while in the constellation of Taurus and is predicted to reach a potential magnitude of 3, which will make it visible to the naked eye.

For further research and information on comets, there are many resources online you can check. One favorite is Comet Watch. It is a UK organization that keeps track of comets. Their website is cometwatch.co.uk.

94. ECLIPSES

Eclipses can be either partial or full. And whether you can see an eclipse as partial or as full will depend on where in North America you live. Here is a photo of a partial solar eclipse.

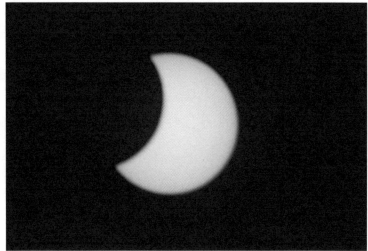

There are two types of eclipses: solar and lunar, both of which involve the moon and the sun. With a solar eclipse, the sun is blotted out by the moon as it passes between the Earth and the sun. With a lunar eclipse, the moon is darkened by the Earth passing between the moon and the sun.

UPCOMING SOLAR ECLIPSES VIEWABLE IN NORTH AMERICA

DATE	TYPE	VISIBLE IN:
June 10, 2021	partial	New England and parts of eastern Canada
October 14, 2023	partial	All of North America
April 8, 2024	total	Portions of North America, but at least partial in all of North America

UPCOMING LUNAR ECLIPSES VIEWABLE IN NORTH AMERICA

DATE	TYPE	VISIBLE IN:
January 20–21, 2019	total	All of North America
May 26, 2021	total	Western United States
November 19, 2021	partial	All of North America
November 8, 2022	total	Western United States
September 17, 2024	partial	Eastern United States

95. TRANSITS

A transit is a celestial event where a smaller body passes in front of a larger body as seen by us here on Earth. They happen frequently in various ways. The most common transits and easiest to see are the transits of the Jovian satellites in front of Jupiter.

With some transits, you cannot see the smaller celestial body. You can only see its shadow on the larger body. In the NASA picture here, you can see the moon of Jupiter, Io, and you can also see its shadow on the surface of Jupiter.

Other examples of transits include Mercury and Venus passing in front of the Sun. Mercury will make a transit on November 11, 2017, but the next transit of Venus will not occur until December 2117.

96. OCCULTATIONS

An occultation is when a celestial object is eclipsed by the moon or another solar system body. The moon, as it travels around the Earth, will occult various objects, including stars and planets. These occultations are predictable and they happen often, but the majority of occultations happen with the moon and very dim celestial objects like low-magnitude stars. However, there are some that are extremely interesting to watch, like when the moon occults a planet or a bright star. Here are some of the more notable upcoming lunar occultations.

- The moon and Venus: January 31, 2019
- The moon and Tejat, a bright star in Gemini: Several occurrences at the end of 2018 and throughout 2019
- The moon and Saturn: Several occultations occurring between April and November of 2019

97. KEMBLE'S CASCADE (NGC 1502)

OBJECT TYPE: Open cluster and optical alignment

CONSTELLATION: Camelopardalis

COORDINATES: 04h 07m 48s, +62° 20' 00"

MAGNITUDES: The brightest star in the cascade is magnitude 5. The rest of the stars average a magnitude of 9.

SEASON: Available year-round for most of North America

DIFFICULTY: Easy

This is a unique formation of stars. There is nothing else like it in the night sky. It is an accidental optical alignment of twenty stars that form a straight line. And none of the stars are related to each other. The whole formation of stars is about six times the width of the full moon. One end of the cascade starts at a cluster of stars named NGC 1502; the stars appear to pour into the cluster.

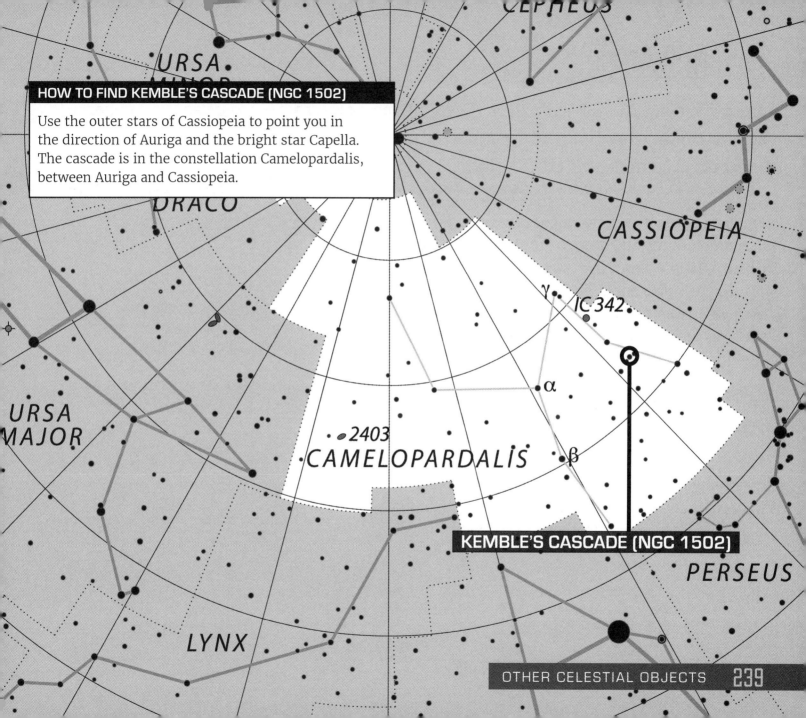

HOW TO FIND KEMBLE'S CASCADE (NGC 1502)

Use the outer stars of Cassiopeia to point you in the direction of Auriga and the bright star Capella. The cascade is in the constellation Camelopardalis, between Auriga and Cassiopeia.

CEPHEUS

URSA MINOR

DRACO

CASSIOPEIA

γ

IC 342

α

URSA MAJOR

2403

β

CAMELOPARDALIS

KEMBLE'S CASCADE (NGC 1502)

PERSEUS

LYNX

98. GLIESE 581

OBJECT TYPE: Red dwarf star

CONSTELLATION: Libra

APPARENT MAGNITUDE: +10

COORDINATES: 15h 19m 26.83s, -07° 43' 20.20"

SEASON: June through September

DIFFICULTY: Medium

Finding planets that circle around other stars is now a common thing, and over 200 of them are known. Gliese 581 is a red dwarf star circled by several planets. It is also one of the more famous systems because it was initially believed to have a planet that was of the right size and in the right orbit for the potential of liquid water, which means there was a higher probability of life. This has, in recent years, been disproved, but Gliese 581 nonetheless maintains some of its fame.

The planets cannot be seen but the star is a red dwarf around magnitude 10 and only about twenty light years away. There is a bit of a profound feeling that goes with viewing another solar system.

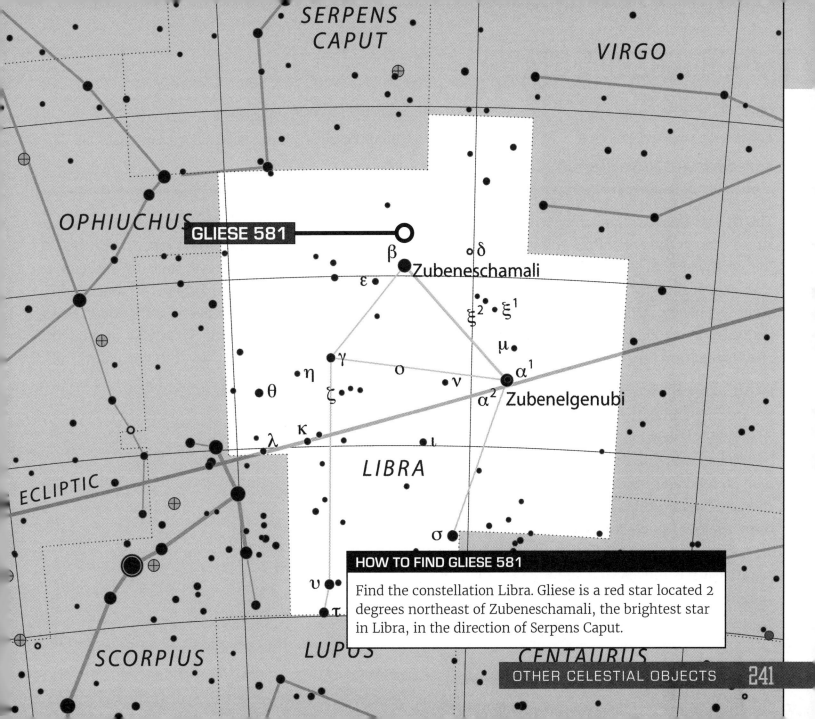

SERPENS
CAPUT

VIRGO

OPHIUCHUS

GLIESE 581

β
Zubeneschamali

δ

ε

ξ² ξ¹

μ

α¹

γ

o

ν

α² Zubenelgenubi

η

θ

ζ

κ

λ

ι

LIBRA

σ

HOW TO FIND GLIESE 581

Find the constellation Libra. Gliese is a red star located 2 degrees northeast of Zubeneschamali, the brightest star in Libra, in the direction of Serpens Caput.

υ

τ

ECLIPTIC

SCORPIUS

LUPUS

CENTAURUS

99. BARNARD'S STAR

OBJECT TYPE: Red dwarf star

APPARENT MAGNITUDE: +9.6

COORDINATES: 17h 57m 48.49s, +04° 41' 36.20"

CONSTELLATION: Ophiuchus

SEASON: June through September; best in July

DIFFICULTY: Difficult

The closest star to our solar system, other than the sun, is Alpha Centauri. But, you can't see it from the northern hemisphere. To see a very close star from North America, you should look to Barnard's Star. Technically, it is the fourth closest to us here on Earth, at six light years away, but it is the only one visible from the northern hemisphere.

Because it is so close to us, and both it and our sun are moving through the galaxy, this star is slowly changing its location in the sky. This change can be tracked over the course of several years. If you plan on following stargazing or astronomy over the course of years, this star will make a great long-term project for you. Observe it and draw out a star chart with it. Several years from now, you will see that it has moved to a slightly new location in comparison to the background stars.

Finding Barnard's Star is a bit of a challenge because it is almost magnitude ten. You can't first locate it with the naked eye, and it's difficult to quickly differentiate it from other stars. But, finding Barnard's Star is a terrific exercise for you in using your telescope and finding your way around the night sky. I've given you several star charts here that help you zoom farther and farther in to find this star.

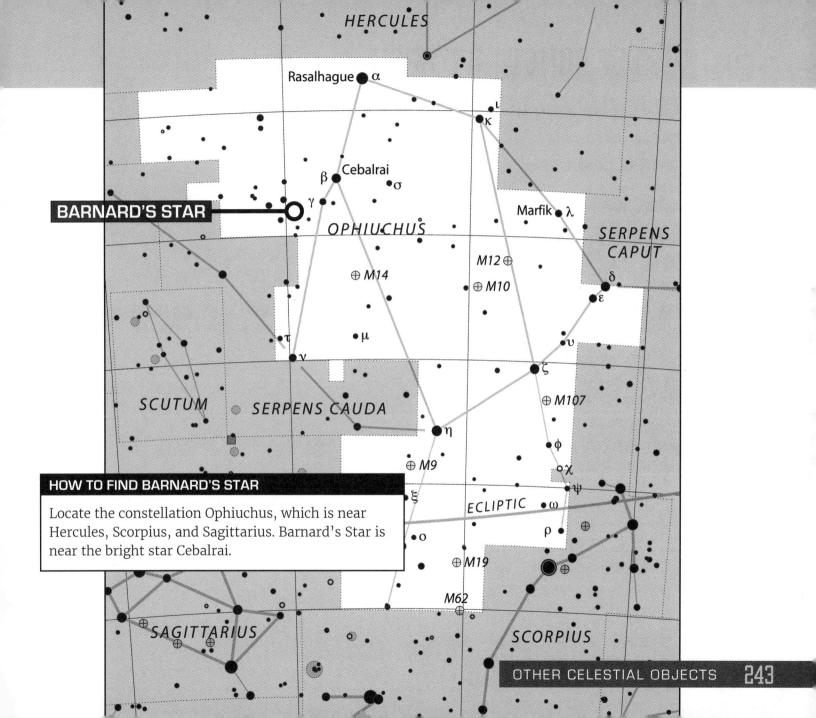

HERCULES

Rasalhague ● α

ι

κ

β ● Cebalrai

σ

BARNARD'S STAR ○ γ

Marfik ● λ

SERPENS
CAPUT

OPHIUCHUS

M12 ⊕

⊕ M14

⊕ M10

δ

ε

τ

μ

υ

ν

ζ

SCUTUM   SERPENS CAUDA

⊕ M107

η

φ

⊕ M9

χ

ξ

ψ

ECLIPTIC   ● ω

ρ

o

⊕ M19

M62

SAGITTARIUS

SCORPIUS

HOW TO FIND BARNARD'S STAR

Locate the constellation Ophiuchus, which is near
Hercules, Scorpius, and Sagittarius. Barnard's Star is
near the bright star Cebalrai.

100. A FALSE COMET (NGC 6231)

OBJECT TYPE: Globular cluster and optical alignment

CONSTELLATION: Scorpius

APPARENT MAGNITUDE OF NGC 6231: 2.6

COORDINATES OF NGC 6231:
16h 54m 00s, -41° 48' 00"

SEASON: June through September; best in July

DIFFICULTY: Difficult

This star cluster is an interesting and random orientation of stars. The globular cluster, NGC 6231, looks like the ball head of a comet. From this cluster, a series of stars form a curved shape, giving the appearance of the comet's tail. The False Comet is far south in the constellation of Scorpius, so it may not be available to you if you are in the northern part of the United States. Typically, it is best viewed with the naked eye because it does span a large area, but you can also give it a try with binoculars or your lowest power eyepiece.

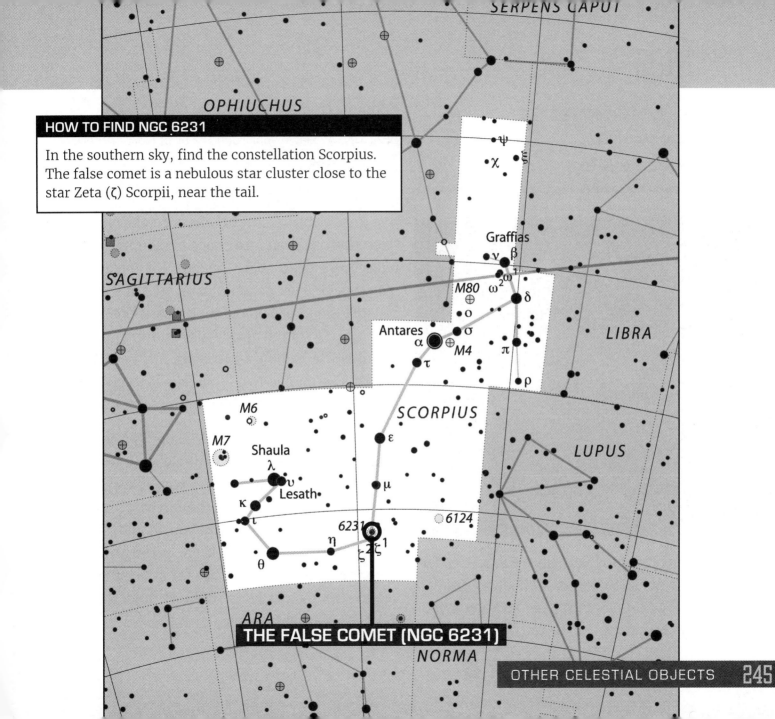

HOW TO FIND NGC 6231

In the southern sky, find the constellation Scorpius. The false comet is a nebulous star cluster close to the star Zeta (ζ) Scorpii, near the tail.

SERPENS CAPUT

OPHIUCHUS

SAGITTARIUS

ψ
χ • ξ

Graffias
ν β
ω
M80 ω²
δ

Antares
α σ
M4 π
τ
ρ

LIBRA

M6

M7

SCORPIUS
ε

Shaula
λ
υ
Lesath
κ
ι
θ

μ

6124

6231
η
ζ2 ζ1

LUPUS

ARA

THE FALSE COMET (NGC 6231)

NORMA

101. M109

OBJECT TYPE: Barred spiral galaxy

CONSTELLATION: Ursa Major

APPARENT MAGNITUDE: +10.6

COORDINATES: 11h 57.6m 00s, +53° 23' 00"

SEASON: Available year round to all of North America

DIFFICULTY: Medium

What is the farthest thing you can see in space? The answer to this question varies a bit depending on your telescope! Of course, the bigger and better your telescope, the farther into space you can see with it.

Just to give you a frame of reference, the farthest any telescope has seen into space is seen in the Hubble telescope's deep space image. NASA pointed that telescope at a very small, blank area of space in the constellation Fornax and took a series of exposures over a period of ten days. This resulting photograph is called the Hubble Ultra Deep Field (HUDF). The image is estimated to contain about 10,000 galaxies and it peers about 13 billion light years away.

Typically, the Andromeda Galaxy is the farthest thing you can see with your naked eye. It is 2.5 million light years away. With your telescope, M109 is your best shot at seeing the farthest and deepest into space. When you find it, you should ponder the thought that the light you are seeing from it has been traveling through space for at least 60 million years—the galaxy is between 60 and 107 million light years away from us.

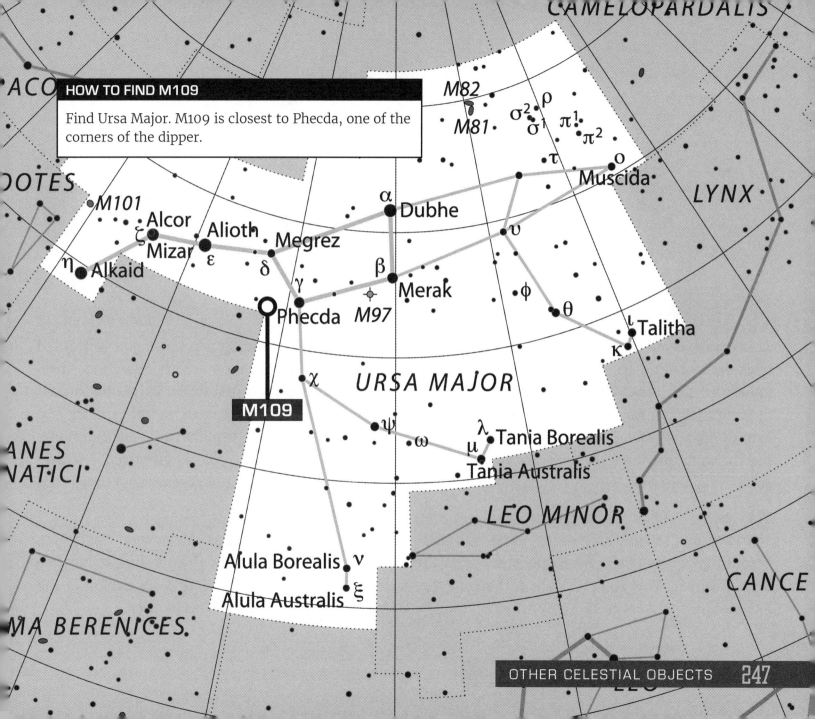

CAMELOPARDALIS

HOW TO FIND M109

Find Ursa Major. M109 is closest to Phecda, one of the corners of the dipper.

M82

M81

σ² ρ
σ¹ π¹
π²

τ

o
Muscida

LYNX

M101

Alcor
ζ
Mizar
Alioth
ε
Megrez
δ
γ
Phecda

α
Dubhe

υ

η
Alkaid

β
Merak
M97

φ

θ

ι Talitha
κ

M109

χ

URSA MAJOR

ψ
ω

λ
μ
Tania Borealis
Tania Australis

LEO MINOR

ANES
NATICI

Alula Borealis
ν
Alula Australis
ξ

CANCE

MA BERENICES

CONCLUSION

This book is all about one thing—getting the telescope out of the garage or the basement and out under the night sky! Over the course of forty years of stargazing, I have noticed a pattern. People love astronomy and are very curious about it, so to feed the curiosity, many will buy a telescope. But then, something happens!

They take it out at night, give it a try, get a bit confused, can't figure out the star charts, and well... before they know it, the telescope is in the garage collecting dust. This book is all about changing that experience.

I hope that from reading this book, you've learned how to get a rewarding and rich experience from the night sky and your telescope. This book showed you how to use a telescope to its full capacity and how to find things in the night sky. Armed with this knowledge, you will be able to get a sense of awe and wonder, just like I have gotten many times with my telescope.

This book is a project that I have been wanting to undertake for a long time. I hope that it inspires you to spend time under dark skies exploring the wonders that our universe has to offer.

RESOURCES

There are lot of excellent programs and websites that will help you learn more about how to use a telescope and how to find objects in the night sky. Here are some of my favorites.

TelescopeNerd.com. This is my website, with lots of resources on telescopes, astronomy, and tutorials on how to make your own telescope.

Stellarium.org. This is a free, open-source planetarium software that you download and install on your computer. With it, you can observe the night sky from any location on Earth, on any day, at any time, in the future or the past. It has a large array of objects in its database, including meteor showers, planets, the sun, the moon, and just about anything else you could want to see in the night sky. With planetarium software like this, the most useful function is the ability to see exactly what is available tonight and where it is located in the sky. This is a highly recommended, and free, program.

3towers.com. The website of the Grasslands Observatory, in southeast Arizona. Visit their website for a lot of great photographs of the Messier objects and other celestial objects, including Barnard Objects, planets, and much more. Their website is a great example of what amateur astronomers can do. They were kind enough to allow me to use many of their Messier object photographs in this book.

SkyAndTelescope.com. This is an excellent professionally run website that is the online companion to their magazine. Just about everything you could want is on their website, including star charts, lunar calendars, and more. They also have a great list of local astronomy clubs and groups; if you're looking to join one, visit www.skyandtelescope .com/astronomy-clubs-organizations.

WorldWideTelescope.org. An online planetarium by the American Astronomical Society. You can also download it and use it on your desktop.

GoogleSky (google.com/sky). The google maps version of the night sky. This is a wonderful graphic planetarium program with lots of rich imagery.

CometWatch.co.uk. If you are interested in hunting comets, this website keeps an updated list on them, when they occur, and where you can find them.

Deepskystacker.free.fr. This is a free software for stacking astrophotos. This is where the software combines a series of digital photographs together to make a single photograph that is brighter and more detailed; it also controls and removes noise.

Stark–labs.com/phdguiding.html. This software is for taking long exposure photographs with an autoguider.

GLOSSARY

Afocal photography: A method of taking astrophotos by holding a camera up against the eyepiece of the telescope. You put the camera at the focal point of the telescope.

Apparent magnitude: How bright a star or other celestial object appears to us here on Earth.

Arc minutes and arc seconds: Angles are used to measure the location of objects in space. The whole of the celestial sphere is measured in 360 degrees as a circle. Each one of these degrees is divided into a smaller unit of 60 arc minutes, and each of these arc minutes is divided into a smaller unit of 60 arc seconds.

Asterism: A visual pattern that is created by unrelated stars. To our eyes they appear to be related because they form a pattern, but in space they are often very far away from each other. The Pleiades are a good example of a famous asterism.

Averted vision: A technique that helps you see faint objects in the telescope more clearly and with more detail. You do it by not looking directly at the object. You look a little to the side of it while still concentrating on seeing the object. This technique brings into play more sensitive areas of your eye.

Binary star: A pair of stars very close together.

Catadioptric telescope: A type of telescope that uses both lenses and mirrors in its optical assembly.

Circumpolar constellations: These are the constellations nearest to the North Pole, such as Ursa Major, Ursa Minor, Cassiopeia, Draco, and Cepheus.

Comet: A ball of ice and dust in an elliptical orbit around the sun. It can develop a tail as it approaches the sun. This tail always points away from the sun.

Constellation: A grouping of stars that form a familiar or recognizable shape. There are eighty-eight officially recognized constellations that cover the complete night sky.

Dark nebula: An area of space full of dark dust and gas that isn't illuminated. It blocks out light from stars behind it and is often overlooked as simply an empty area of space.

Declination: A measure of distance on the celestial sphere. It is measured north or south from the celestial equator.

Diffuse nebula: Wide areas of space containing interstellar matter made up of gas and dust. The larger diffuse nebulae are often known as stellar nurseries, as they are in the process of creating new stars.

Dobsonian mount: A mount style for reflector telescopes. It is a simplified and very easy to use mounting style.

Eclipse: A phenomenon during which one celestial body moves into a position between two other celestial bodies, casting a shadow. The most well-known type of eclipse is a solar eclipse, when the moon eclipses the sun. Another common type of eclipse is a lunar eclipse, where the Earth passes between the moon and the sun. This causes the shadow of the Earth to darken the moon.

Ecliptic: A narrow band through the sky that the planets, moon, and sun travel in.

Elliptical cluster: A cluster of stars that has an oblong shape rather than a circular one.

Eyepiece: A small cylindrical object with a lens or series of lenses inside. It is called an eyepiece because it is where you directly put your eye when looking through a telescope.

Eyepiece projection photography: A method of taking photographs with a camera where the camera is attached to a telescope. The camera has no lens; instead, the eyepiece in the telescope is used.

Finder scope: This is a smaller telescope mounted on the main telescope. It is an important tool for helping you to find objects in the night sky.

Focal length: The distance of the light path of a telescope or eyepiece. This distance is measured from the point where the optics begin bending light to the point where the light is brought to a focus. Focal length is important in a telescope and in an eyepiece.

Gibbous: This term is generally used in reference to the moon when it is nearly, but not completely, full. The moon is generally around three quarters full when referred to as gibbous.

Globular cluster: A cluster of stars that form a uniform globular shape. The number of stars can range from several hundred to several million.

GoTo telescope: A computerized telescope that will automatically "go to" a celestial object when you enter the coordinates or object name.

Lunar eclipse: A type of eclipse where the Earth passes between the sun and the moon, and casts a visible shadow on the moon.

Messier objects: A list of 110 astronomical objects compiled by the astronomer Charles Messier. These objects are predominantly galaxies, nebulae, and star clusters.

Meteor: Also known as a shooting star or a falling star. It is a small particle of dust, debris, or metal that enters Earth's atmosphere and burns up in a trail.

Meteorite: Most meteors burn up completely as they enter Earth's atmosphere. If a portion of the meteor makes it all the way to the Earth's surface, that surviving fragment is a meteorite.

Nebula: An interstellar cloud of gases and dust. It can be of many different colors and of enormous size.

Newtonian telescope: A type of telescope invented by Sir Isaac Newton. It uses a parabolic mirror as its primary means of gathering light.

NGC (New General Catalog): A catalog of deep space objects including nebulae, star clusters, and galaxies. It contains 7,840 objects referred to as the NGC objects.

Nova: A star that temporarily increases its brightness significantly, sometimes by as much as 100,000 times its original brightness. It maintains this brightness for several months, then returns to its original brightness.

Occultation: When a celestial object occludes another object. The most common type of occultation occurs when the moon passes in front of a star or planet. The star or planet disappears behind the moon for a period of time then reappears on the other side.

Open cluster: A loose configuration of stars that form a group. This group can be gravitationally bound to each other, forming a system, or they can simply appear to be a cluster due to our line of sight from Earth.

Optics: The parts of a telescope that manipulate and magnify light. It can be a system of lenses or mirrors or a combination of both.

Piggybacking: The method of mounting a camera to the top of a telescope, enabling the camera to take astrophotos.

Planetary nebula: A type of nebula that resembles the shape and look of a planet.

Planisphere: A rotating star chart that shows the orientation of the stars and constellations for any given day and hour.

Prime focus photography: A style of astrophotography where a camera is added to the telescope. There is no eyepiece on the telescope and no lens on the camera. The telescope acts as the lens for the camera.

Reflector telescope: A type of telescope that uses a parabolic mirror as its primary source of light gathering.

Refractor telescope: A type of telescope that uses a lens as its primary source of light gathering.

Right ascension: A measure of distance on the celestial sphere (the sky). It is measured east and west. Think of it as a measure that goes in the direction of the sun as in from sunrise to sunset. If you face north the sun will "ascend" from the eastern horizon.

Solar eclipse: A type of eclipse where the moon passes between the Earth and the sun, and causes the sun to appear to be blacked out.

Stacking: A method of taking multiple short-exposure photographs of celestial objects then stacking them together with a software program to create a brighter and better image.

Star cloud: An extremely dense accumulation of stars, numbering in the millions. This extreme density gives it the appearance of a cloud.

Supermassive black hole: The largest type of black hole. Its mass can equal millions or even billions of solar masses. A supermassive black hole can be found at the center of almost all known massive galaxies, including our own Milky Way.

Supernova: Massive stars come to the end of their life cycle with an explosion of extreme energy, increasing their brightness by as much as 20 magnitudes.

Supernova remnant nebula: The gas, dust, debris, and ejected material left over from a supernova explosion.

Totality: A term for when the moon completely blocks out the sun.

Transit: A celestial event where a small body passes in front of a larger body. A common transit would be when a moon of Jupiter passes in front of Jupiter.

UDCA (Universal Digital Camera Adapter): A mechanical adapter that allows you to connect a digital camera to the eyepiece of a telescope.

Variable star: A star that changes in apparent magnitude. This can occur because the star waxes and wanes in strength, or because the star is regularly eclipsed by another object, such as a large planet or another star.

Zenith: A spot directly overhead in the sky.

PHOTO CREDITS

ABOUT THE AUTHOR

Will Kalif is a writer, webmaster, and an avid telescope enthusiast currently living in New England. He worked a full career in a variety of fields including robotics, computers, and electronics. He has been passionate about telescopes and the wonders of the night sky ever since he received his first telescope as a teenager. For several decades now, he has been making and using his own telescopes and helping other people enjoy the various things that can be seen on a dark and starry night. You can learn more about telescopes, telescope making, and astronomy by visiting his website at telescopenerd.com.